Engineering and Society:
Working Towards Social Justice

Synthesis Lectures on Engineers, Technology and Society

Editor
Caroline Baillie, *University of Western Australia*

Engineering and Society: Working Towards Social Justice
Caroline Baillie and George Catalano
2009

Engineering: Women and Leadership
Corri Zoli, Shobha Bhatia, Valerie Davidson, Kelly Rusch
2008

Bridging the Gap Between Engineering and the Global World:
A Case Study of the Coconut (Coir) Fiber Industry in Kerala, India
Shobha K. Bhatia, Jennifer L. Smith
2008

Engineering and Social Justice
Donna Riley
2008

Engineering, Poverty, and the Earth
George D. Catalano
2007

Engineers within a Local and Global Society
Caroline Baillie
2006

Globalization, Engineering, and Creativity
John Reader
2006

Engineering Ethics: Peace, Justice, and the Earth
George D. Catalano
2006

Engineering and Society: Working Towards Social Justice

Caroline Baillie and George Catalano

www.morganclaypool.com

ISBN: 9781598296624   paperback
A Publication in the Morgan & Claypool Publishers series
SYNTHESIS LECTURES ON ENGINEERS, TECHNOLOGY AND SOCIETY

Series Editor: Caroline Baillie, *University of Western Australia*

Series ISSN
Synthesis Lectures on Engineers, Technology and Society
Print 1933-3633   Electronic 1933-3461

Engineering and Society: Working Towards Social Justice

Caroline Baillie
University of Western Australia

George Catalano
State University of New York at Binghamton

SYNTHESIS LECTURES ON ENGINEERS, TECHNOLOGY AND SOCIETY

ABSTRACT

Engineers work in an increasingly complex entanglement of ideas, people, cultures, technology, systems and environments. Today, decisions made by engineers often have serious implications for not only their clients but for society as a whole and the natural world. Such decisions may potentially influence cultures, ways of living, as well as alter ecosystems which are in delicate balance. In order to make appropriate decisions and to co-create ideas and innovations within and among the complex networks of communities which currently exist and are be shaped by our decisions, we need to regain our place as professionals, to realise the significance of our work and to take responsibility in a much deeper sense. Engineers must develop the 'ability to respond' to emerging needs of all people, across all cultures. To do this requires insights and knowledge which are at present largely within the domain of the social and political sciences but which needs to be shared with our students in ways which are meaningful and relevant to engineering. This book attempts to do just that. In Part 1 Baillie introduces ideas associated with the ways in which engineers relate to the communities in which they work. Drawing on scholarship from science and technology studies, globalisation and development studies, as well as work in science communication and dialogue, this introductory text sets the scene for an engineering community which engages with the public. In Part 2 Catalano frames the thinking processes necessary to create ethical and just decisions in engineering, to understand the implications of our current decision making processes and think about ways in which we might adapt these to become more socially just in the future. In Part 3 Baillie and Catalano have provided case studies of everyday issues such as water, garbage and alarm clocks, to help us consider how we might see through the lenses of our new knowledge from Parts 1 and 2 and apply this to our every day existence as engineers.

KEYWORDS

engineering and society, social justice, ethics, engineering education, globalisation, public dialogue with engineering, engineering development, engineering studies

Contents

Part I

Engineering and Society

Figure 1. Abandoned fishermen's houses near Bonavista, Newfoundland, ten years after the cod moratorium. Overfishing, enabled by modern technologies, caused the fish stocks to recede to dangerously low levels.

C H A P T E R   1

Introduction

1.1   INITIAL THOUGHTS

Both of the authors of these volumes are on a mission. We won't try to hide this fact. We are working on a pathway towards engineering, which is increasingly just for all people in the world, as well as environmentally sustainable. In this book, we present many issues which might relate to this context and we want to ask you, the reader, to question whether it is possible to define a 'good,' 'right' or 'just' engineer. In order to do this, we need theoretical underpinning as well as practical realities. We need to look historically at what engineering has been and consider what it could be/ should be/ what is its function? We need to consider the notion of 'engineering' in a society and what that implies. These three volumes will allow you to walk through some of the relevant concepts from disciplines which seem very far from engineering, but which are critical to draw from if we are to prepare engineers to contribute to an increasingly just society. We, therefore, draw on different aspects of philosophy, psychology, economics, development studies, politics, history and sociology in our aim to understand our social context. In the text which follows, we present a range of views, including our own. We freely admit our own positions and ask you to critique us as well as the other authors we discuss. We do not ask you to accept our way of looking at the world, but to question it, to question others and to question your own views – then to make up your mind and act on your own beliefs.

1.2   WINDOWS ON SOCIETY

Throughout the text we will be taking a critical perspective. This does not mean that we will criticise – but that we will critique through different lenses. 'Critical Theory' is an amalgam of philosophical and social scientific techniques providing a way to systematically, critically and yet constructively analyse systems of thought and practice. It includes a variety of possible approaches and perspectives by which to analyse 'not only cultural artifacts but also their contexts – social, political, historical, gender, ethnic' Sim and Van Loon (2001). Critical theory is a tool which enables us to put engineering 'under the microscope' and to see whether we like what we see or whether we want to change it. In order to understand what we mean at this point, we will give you an example.

Imagine that a large engineering firm is 'downsizing.' The management of the company needs to consider the implications of what they are doing in order to make decisions. Workers are often faced with this situation and remain for many months or years unemployed after this event as they can find no work. This situation involves many complex arguments which we will explore later in the book. However, for now, to give an example of a critical perspective on this situation, we might explore the comparison between the individual versus social meaning of the situation. Osborne and Van Loon (1998) ask us to consider the question that some of these workers might ask, 'Why am I unemployed?' We might consider how much respect an unemployed person has in society. Students of mine have actually commented that 'We can't help it if they don't respond to education.' It seems as if the common sense view is that the individual is responsible. In fact, there are many reasons for the surplus labour which causes people to be laid off which include:

Technological change (new machines)

Changed work practices (efficiency)

Work done in other countries (globalisation)

Political change (government policy)

Cultural change (different products wanted)

Lack of requisite skills (no access to education or retraining)

None of these has anything to do with the individual worker but 'blaming the individual is common political practice' (Osborne and Van Loon, 1998, p10). Students will be asked to consider their own views and to debate and critique the various perspectives presented to them when focusing through 'windows' or lenses on typical areas well known to students. A very brief consideration of these windows will be presented below.

1.3   WINDOWS, DISCIPLINARY PERSPECTIVES, AND THE FLOW OF TIME

Engineering uses mathematical techniques to solve equations which typically arise from modelling. One such mathematical technique, which can be used to model time in a dynamic process, is referred to as the method of characteristics. A visual representation of the method of characteristics can be seen in the flow of water as it passes an obstacle in a stream. Students are asked to imagine they are sitting along a stream's bank and they reach over and place a stick vertically in the flowing water. A wake region, triangular in shape, will form behind the stick and this wake will spread out so that encompasses more and more of the stream. This region is known as the zone of influence—that is, every location downstream of the stick is influenced by the disruption in the flow caused by the stick's presence. The extent of the influence of decisions can be seen to grow in time until they eventually disappear. Suppose through some cosmic quirk, we could cause the flow to turn around and flow upstream. There would be a similar triangular form, referred to as the domain of dependence, extending upstream from the stick. This can be seen as analogous to looking back in time. In total then, we have a moment in time which has been preceded by a series of events and which is followed by a set of consequences.

Perhaps then we can extend the metaphor even further by placing various experts from different disciplines watching the unfolding of events through a window of a nearby building. Each expert will "see" the developing events within the context of a particular discipline or perspective. Students will "see" that there are not events in history, sociology, political sciences, gender studies, economics, and global studies but simply events that can be understood in different ways and in different contexts. Each decision we make at this present moment in time is a result of the decisions and work of countless others at countless other earlier times. Equally as importantly, you will be introduced to

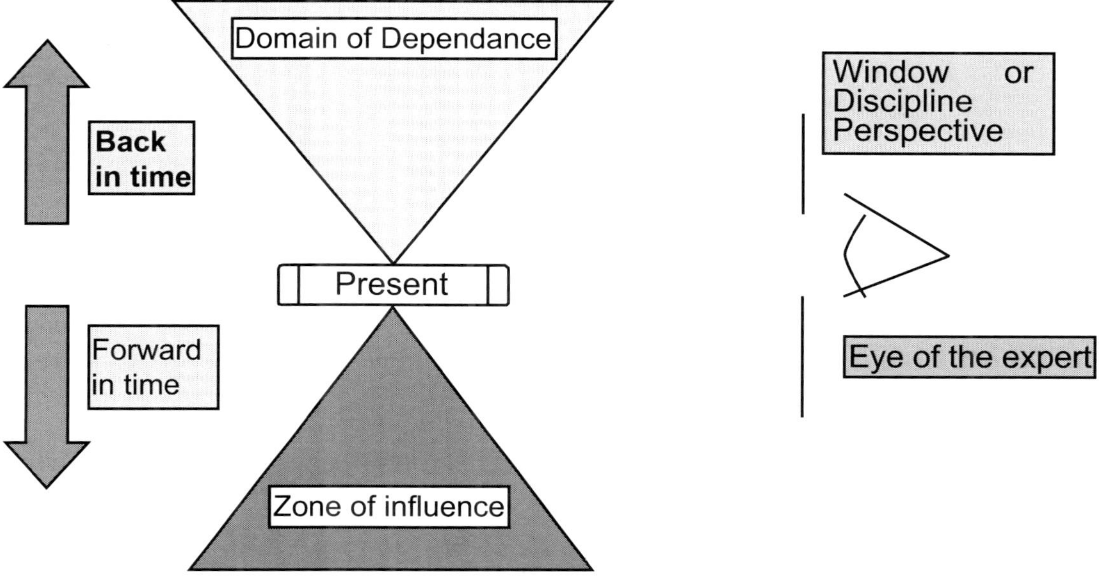

Figure 1.1. Windows of perception.

the notion that each decision you make as a human being and as an engineer will have consequences that may extend far into the future and may be understood by others in a completely different way. Let's reflect on the implications for our "zone of influence" and how we might relate it to our notion of "windows." The point we are trying to make is that what we see, what we come to regard as an objective truth, is a function of what we have experienced in our lives. An event which occurred during our childhood and how it effects our notion of reality is transported downstream much like the water that flows as it passes the obstacle in its path. If this is true for events in our lives then certainly it makes sense to think of how our training and education can influence how we see the world and this influence will continue to be with us as time advances. This idea can certainly be extended for the case of 'experts,' how they see the world and what implications these perceptions might have, which are propagated forward in time.

Lenses or Disciplinary Perspectives

In the chapters that follow, we will not explicitly state each time we use a different lens; but to give you an idea of the different influences used to develop the thinking of this book, we present some of the typical lenses below.

Lessons from Sociology

This perspective frames how engineering is produced and utilized in local and global communities, how different engineering projects originate, how social needs are defined (participatory needs analysis), notions of social welfare and how the engineering impacts and affects different receptor societies. It also considers worker organization, employee relations, labour unions and social capital.

Lessons from History

This perspective deals with an exploration of the basics of the Industrial Revolution and the relationship between automation and labour. We consider the implications of this today with reference to globalisation and how it affects engineering organisations in developing countries. We will also explore the effect of increasing mechanisation on systems of thought and action.

Globalisation Debate

Current issues influenced by the global market will be considered using the work of key texts on globalization - introducing the World Trade Organisation (WTO), 'free trade' global capitalism and sustainability.

Lesson from Economics

Different economic models will draw on a range of perspectives from capitalist, to Marxist viewpoints. Economic development, for example, might be considered from the perspective of Amartya Sen's work on 'freedom.' His approach differentiates itself from traditional practical ethics and economic policy analysis, such as economic concentration on income and wealth (rather than on characteristics of human lives and substantive freedoms), utilitarian focus on mental satisfaction (rather than on creative discontent and constructive dissatisfaction) and the libertarian focus on procedures for liberty (with neglect of consequences of those procedures). There are, of course, connections to low income, but this needs to be integrated into a bigger picture. Poverty then can be seen as deprivation of basic capabilities rather than just low income, e.g., premature mortality, significant undernourishment, persistent morbidity and illiteracy.

Lessons from Political Studies

We will focus on the recent political changes that have occurred and the ways in which these affect the relationship between engineering and society. For example, the massive political-economic change that occurred with the collapse of state socialism, the establishment of private property, the influx of multinational capital into Eastern European countries and the integration of firms into the global economy.

Lessons from Environmental Studies

Recent developments on climate change, life cycle analysis, sustainable development, Kyoto protocol and different governmental perspectives will be explored and links to public health questioned.

Lessons from Psychology

From this perspective, we will consider personal development versus professional development. We will refer to recent work from psychology and Gestalt therapy, which helps the student consider their own blocks to self awareness, an understanding of personal ethics and potential ways of enhancing individual freedom and personal creative potential.

1.4   READING (BAILLIE 1998)

Engineering students, attempting for the first time to read and write in what is usually considered to be more of a social science style, can become frustrated and disillusioned, believing that it is an irrelevant skill to learn. However, learning to learn how to read and write in different ways, so as to adapt in the future, is critical. You never know who you may have to write for in the future and who your audience may be. You might become a policy maker or the director of a grant funding agency. You might become the Vice Chancellor of a university with many disciplines.

We have all been in the situation where we have read through a text and cannot remember any of it. In this case, we need help in concentrating on what we want to know or find out rather than treating each part equally. Alternatively, you may have tried to read something in another discipline and although the words seem to be English and you understand all of them, together they make no sense. Here are some useful ideas:

Selectivity

There is no point in reading for the sake of it unless you are lost in a novel or a piece of beautiful literature. You need to focus on the relevant material.

Strategy

- Survey the whole, question the specifics, read the details, recall periodically during your reading and keep notes, review. Look for the main ideas, realise that you and the author may differ on this. Question assumptions.

- Pick out the details -which are the author's important points?

- Evaluate the text - don't believe all you read without argument or evidence.

  The important features of reading can be summarised by the acronym SQ4R.
  Scan or Survey - for an overview.
  Question - what do you want from the reading?
  Read - look for key material and read selected detail carefully.
  Recall - periodically consider what you have read.
  Review - put it all together.
  Relate - tie in with other topics.

1.5   CREATIVE THINKING

In all engineering work, creativity is necessary; but when it comes to working at the interface of different ways of thinking and being, it becomes essential. The next section looks at some skills necessary to think about creative problem solving.

"His handlebars had started slipping"

"You're going to have to shim those out"

"What's shim?"

"It's a thin, flat strip of metal...I've got some right here"

"I said gleefully, holding up a can of beer in my hand"

"What, the can?"

"I believe now he was actually offended." I had had the nerve to propose" repair of his new eighteen-hundred-dollar (Author's note !!!!!!!1974) BMW, the pride of a half century of German finesse with a piece of old beer can! I was seeing what the shim meant. He was seeing what the shim was. We were both looking at the same thing....from a completely different dimension" (Zen and the Art of Motor Cycle Maintenance, p60).

You need to feel safe to explore - i.e., you know what the boundaries are, what sort of thing is expected of you, what the professor wants from you, is it truly open-ended or are there criteria you must satisfy? You need to allow yourself to think up as many ideas as possible, rejecting none in the first instance. However, there is a huge danger when working with social contexts, to think that coming up with a great solution by yourself or your team is okay. We will come back to the idea of participation in the section on engineering and the public. It is critical to hold some form of needs assessment as a problem definition stage before attempting any problem solving.

Creative Potential

We define creativity as 'shared imagination' (Dewulf and Baillie, 1999). Imagination is novel, rather then visual memory, it is individual or personal. Sharing means a sort of negotiation with the user.

Often in discussions on creativity, the following questions arise: Are there creative types of people who have certain intelligence, knowledge, technical skills, special talents and values? What about environmental conditions – politico-religious factors, cultural factors, socio-economic factors and educational factors? Does a creative person have a particular personality – internal motivation, confidence, non-conformity, good self image, emotional, perceptual and open to new ideas (Boden, Jones in Isaksen et al., 1993)? The literature, in fact, suggests that each individual has a creative potential (Fig. 1.3).

The turquoise box shows the internal aspects of the individual. These may be brought about by psychological conditions. Obviously, the block or obstacle is the converse of the condition leading towards creativity. For instance, a high level of confidence in ability will help and a low level will hinder. The purple box shows the conditions or obstacles which might help or hinder creativity, from an external perspective. These two boxes will influence the central blue box, which is the creative potential of the individual. In this box, the person's ability to communicate their ideas, to see the

Figure 1.2. Creativity is shared imagination.

possibilities in a situation, and to acknowledge assumptions will depend upon their own conditioning and their environment. The yellow box shows the intrinsic motivation for creativity, which may be set up directly by the individual personality in the turquoise box. The red box is the extrinsic motivation, which may be set up directly from the external environment, given in purple.

However, under less advantageous situations where the obstacles have become insurmountable, the creative potential needs to be aided by certain skills, knowledge and awareness, shown in the orange box. These might be the ability to visualise what one wants to do, to think laterally as well as vertically, to associate different sets of ideas and so on. It appears to be possible to develop the creative potential and to take a back route towards the extrinsic or the intrinsic motivation via development of these areas. Even a negative environment, given the right sort of attitude, can be a motivation towards creative change, if the individual sees it as an opportunity for improvement.

The Creative Process

'I call intuition cosmic fishing, you feel a nibble then you've got to hook the fish, after bathing the hook through preparation and generation, and trolling deep waters with incubation it's time to reel in the catch' (Buckminster Fuller).

The creative process has been described in various ways in most creativity research and, generally, comes down to four phases: preparation, generation, incubation and verification.

In the initial stage, preparation, the problem or question is defined, reformulated and redefined, moving from a given to an understanding. The way in which the question is formulated severely influences the solution finding ability. Instead of seeking the right answer, one can ask 'Is this the

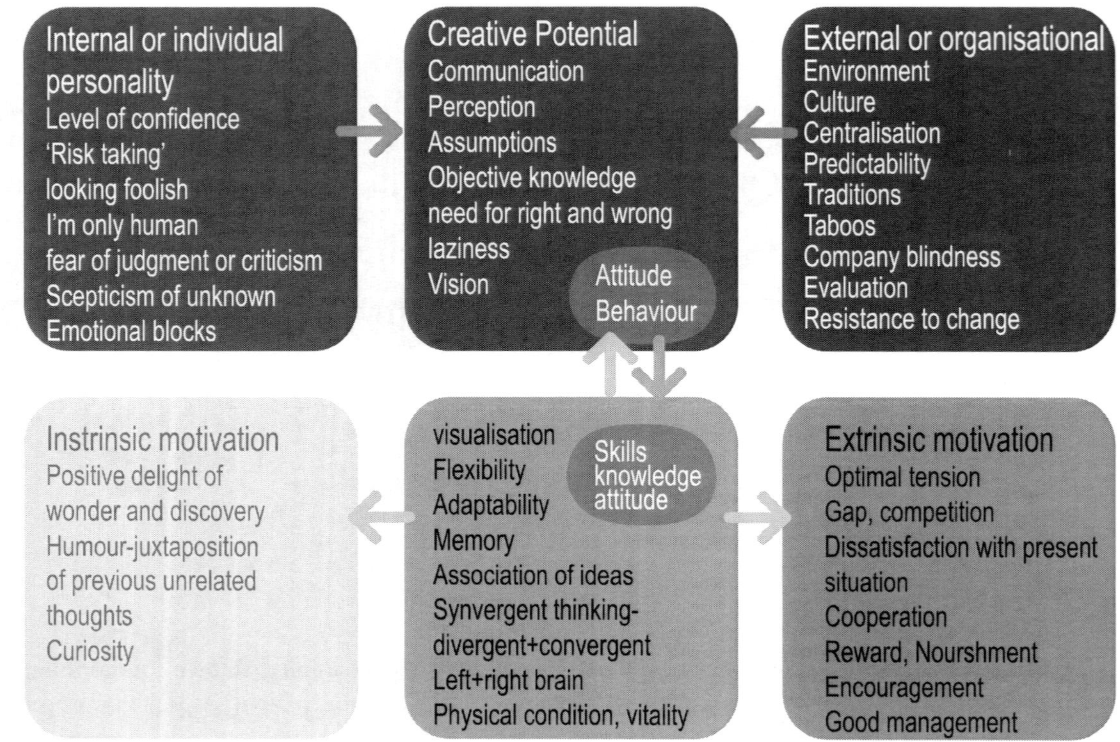

Figure 1.3. The Creative potential.

right question?' Gelb, (1996) points out that our move from a nomadic way of life to settlement occurred by reformulating the question, 'How do we get to the water?' to ' How can we make the water come to us?' This stage is the most critical when working directly with people in social and development contexts.

After this comes the purge. The purge aims to shake out all ideas, even the most obvious ones, as they may hinder the mind from looking further (Osborn in COCD, 1999).

The generation phase involves 'moving beyond habitual pathways of thinking,' (Gelb, 1996) by purging associative concepts to the problem. Osborn described generation as brainstorming (COCD, 1999). Osborn, furthermore, sets up principles for brainstorming: quantity breeds quality; postponement of judgment; hitch-hiking (building on other peoples' ideas); and free-wheeling (continuing unrelated fantasy).

As well as being part of most creative developments, incubation can act on its own, as a subconscious stimuli to an inventive concept. Studies have shown that individuals frequently generate a potential idea after a certain time of incubation (sleep, shower, biking, sailing, etc.), a period of full relaxation or relaxed attention (Tomic and Brouwers, 1998). After incubation, all the ideas are analysed, clustered and evaluated. Very creative people do all of these stages in any order and without thinking. Most of us have to force ourselves to remember the different stages if we are to be as creative and yet appropriate as possible.

1.6   QUESTIONING AND LISTENING

Asking questions

One of the key ways to start to understand different ways of thinking is to question yourself. It might sound obvious; but by taking nothing for granted and by questioning assumptions, you can start from scratch with any person in any situation, so long as you are not transgressing cultural forms of respect by doing so.

Why do I want to know? (What is your purpose in finding out?)
What do I want to know? (What interests you?)
How do I know that? (Where did you get that information?)
Who is affected?
When will this happen?

Listening

'Sit down both of you and don't speak a word till I've finished' (said the Mock Turtle). So they sat down and nobody spoke for some minutes. Alice thought to herself, 'I don't see how he can ever finish if he doesn't begin' (Alice in Wonderland).

We've all been in a situation where we were talking to somebody for several minutes and then realised that we hadn't 'heard' a single word of what the other person was saying when they asked us for a response. We are, in fact, hearing but not listening. We can think faster than the other person can talk, and often we are thinking about what we are going to say next. This becomes even harder when we are talking to people from other contexts or cultures than our own. Accents and cultural norms may affect the communication when language barriers have been removed.

Active listening in a discussion with fellow students/staff/future colleagues is about awareness. Some questions to ask yourself:
Do you always give the other person your version without trying to understand theirs?
Do you talk about yourself most of the time?
Do you change the subject of the conversation when you feel uncomfortable?
Do you try to advise other people, or do you help then come up with their own answers?
Do you acknowledge their feelings or correct them?
Do you allow silences?

1.7   TAKING A POSITION VS. STATING AN OPINION

The most important thing to learn when working at the boundary of disciplines and with different people and cultures is the way of thinking that underlies the actions. A critical theory is one which usually has a socially progressive approach, such as feminist or environmentalist and it is the lens through which you view the reading or writing. You critique it through this lens. We might call this taking a position. You don't have to stick with this position; but when reading a text, it helps when analysing what the author means if they have stated their position when writing it. The following section by Richard Day, Professor of Sociology at Queens University demonstrates this well.

While stating opinions is of some personal value, and is important politically and socially, in an academic context stating opinions is less important than taking positions.

What's the difference?

An opinion is not necessarily informed by any knowledge of the matters upon which one is opining. Anyone can have an opinion about anything. Opinions cannot be refuted, i.e., they can contradict other opinions without difficulty since there is no shared basis for discriminating between them.

Taking a position, on the other hand, means having at least some knowledge of that about which one is speaking, and especially of what others have said in the past, and are saying now. We could say that taking a position means precisely showing that one knows what other positions have been, are being, and could be taken. This shared background is what makes it possible for positions to be compared, contrasted, and evaluated.

The difference can, perhaps, be demonstrated by showing how different kinds of statements work. Here is a statement of an opinion:

"Everyone needs a job so they can buy things."

This is 'common sense,' so you'd think no one could argue with it. However, it is a naive statement in a number of ways:

- it takes for granted the existence of a capitalist economy, in which individuals and communities are separated from the means of meeting their needs directly, and thus are forced to go through the mediation of corporations and markets.
- even within capitalism, it ignores the many possible ways in which one can meet one's needs outside of the money economy, e.g., through delinking, local barter systems, and so on.
- it reinforces capitalist individualism and consumerism.
- it fails to consider the needs of those who cannot work for reasons beyond their control.

To clear out some space in which a position might be taken, instead of an opinion expressed, one would have to address the issues raised above in relation to the social, historical, political-economic construction of 'the job.'

One might do this with reference to Marxist critiques of capitalist political economy; feminist critiques of the marginalization of 'women's work;' postcolonial critiques of the global division of labour and colonialism; anarchist critiques of the addiction to work beyond that which is necessary for sustenance.

In taking a position, one must be careful to always spell out who is saying what, e.g.,

"According to Malcolm X, "The earth's most expensive and pernicious evil is racism." As a Mohawk woman, I can understand why he feels this, since I have also experienced various forms of racism at the hands of European settlers. I would only add that, from a feminist perspective, patriarchy must also be included in any analysis of white racism, since it creates similar effects of oppression within racialized communities."

Here, the observation of Malcolm X is attributed to him; the writer identifies herself, i.e., her positioning as an indigenous woman, and shows how this positioning relates to her interpretation of X's comment. She also identifies her critique of X as coming from a feminist perspective, thus placing her statement in relation to this tradition of theory and practice. There is no doubt who is saying what, or why they are saying it.

1.8 FINAL THOUGHTS

Throughout the texts, we will be challenging the reader with some difficult thoughts. These are not easy questions to answer and there are very few 'right answers.' We hope that by the end of the book you will have started to think in a critical way and be able to take a position on a particular topic and argue for that position from an informed perspective.

Now that you have started to look at the world through different lenses, we present a series of cases of everyday activities and services for you to view. We ask that you view these as a citizen, as an engineer, but also for perhaps the first time, with a view for improving the social justice in the world.

Challenge Box: What is your current position on the relationship between engineering and society? Do you think it's a useful topic to study or one which should be left to the social scientists to worry about?

CHAPTER 2

Engineering and Society

2.1   ENGINEERING IS NOT IN A VACUUM

Technology, and the engineering practice which puts this into place, is not separated from the rest of society, in some sort of vacuum.

Figure 2.1. Engineering is not in a vacuum.

What we imagine, what we believe to be important to develop for progression in our society, will determine what we actually do create. Different values in a society could lead to a completely different set of technological developments.

In this chapter we will explore some recent ideas about the relationship between science, technology, engineering and society.

The objective nature of science (i.e., that there is one truth that is out there waiting to be discovered and, therefore, all technology is inevitable) has been challenged time and again (Kuhn, 1962, Shapin, 1995; Zuckerman, 1988) and technologies are seen to be integrated into existing systems of social relations and social organization. In other words, what scientists do is integrated into a complex mix of economic, social and political issues.

Systems of social organisation both determine and are shaped by what is technically possible. As such, they need be subject to the same critique and reflection that we give to the larger systems of social relations and organizations (Mumford, 1963, 1967; Muller, 1970; Noble, 1977, 1984; MacKenzie and Wajcman, 1985). The play 'Copenhagen,' by Michael Frayn (1998), deals in an interesting way with the issue of the question of 'inevitability' related to scientific progress. The play deals with two famous physicists, Niels Bohr (working for Denmark and the Allies) and Werner Heisenberg, (working for Germany) who were developing knowledge about the atomic bomb, effectively turning science into a technology that would have enormous repercussions. Apart from its devastating effect in World War II, it is still the largest cause of political unrest in the world today. The play deals with a meeting between the two old friends that was known to take place in Denmark during the occupation. The questions raised are: Why did Heisenberg visit Bohr? Did he go there to find out how to make the bomb? To warn his friend that the Germans were making the bomb? Or did he go there to ask his friend and mentor's advice – should they both, in fact, stop the programme, knowing the potential consequences? Could it have been stopped? The suggestion made is that, in fact, the funding for the programme was so small in Germany that they could never build the bomb. Was that the real reason? Audiences, lay public, are amazed that such significant developments are left up to the egos of individual scientists. We might say that it would have been developed somehow, by someone, but this is a question really worth asking: If there is no funding and no support for such a development, is it, in fact, inevitable? Research does not get done in areas which are not funded. Developments cannot be made and technologies turned into commercial products if no-one supports the idea.

Other arguments suggest that technological progress has a sort of life of its own, that it is not developed for a particular purpose, but it does affect the way we live. Technological determinists believe that we must question the impact of science on society and take action if the consequences are not what we would want.

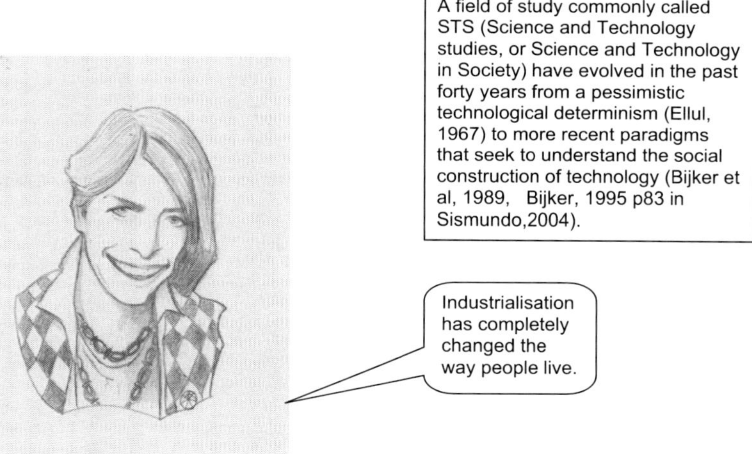

Figure 2.2. Heather Menzies (1989, 1996).

Figure 2.3. Society shapes science and technology.

Challenge Box: Think of five ways in which technology influences the way you live. For each technology, what do you think the key drivers were for its introduction?

Figure 2.4. Science and technology shape society.

2.2   SCIENCE, TECHNOLOGY, AND ENGINEERING

Many sociologists of science and technology spend a good deal of time deciding whether technology is applied science or whether most technological developments occur regardless of basic science (Sismundo, 2004). It also appears to be assumed that those that 'do technology' are engineers. Bijker (Bijker et al., 1989) refers to engineer-sociologists in order to explain the way in which engineers relate to society and the market and we begin to realize that in order to go beyond the mere craft of the technique - engineering requires a broader network of actors and systems, which we will refer to below. Engineering then, might be considered to be a 'technological system.' Technology includes what people know as well as what they do. Suffice to say that pure science, applied science or technology and technological systems (which apply this to a larger scale than simply crafts) or engineering, are often taken as one and the same in social studies. As much as the boundaries are sometimes superfluous it helps us to analyse the links to society better if we are clear on which of these we are referring to. Scientists may be reluctant to admit to their having a responsibility about the potential end use of their discoveries. Engineers on the other hand are very clear about this (or should be) and it is an important part of an accreditation system to consider the ethics of their actions. However, once we see how complex the picture is regarding the interrelationship of societal needs, political drivers and government run funding agencies, market driven demands and corporate funded research, we see that there is rarely any 'pure' science today. It is simply not possible to get funding for expensive equipment and personnel to do research on basic science, which is not considered to be at least potentially capable of enhancing the livelihoods of the citizens of the country supporting the funding scheme. Thus, it is the very decision about what constitutes enhancement in lifestyle that makes even pure science, political.

Lets look at the three main ways of considering the interaction of science, technology and society.

2.3   SCIENCE AND TECHNOLOGY SHAPE SOCIETY

This has been considered to be both good and bad by different authors over the last two centuries. Some of the chief supporters of this perspective have worked on the principle that science is strongly related to the economy of the country. Science and technology are seen as positive forces of socioeconomic change by many in the 1930's onwards. Bernal, for example, argued for science as a cornerstone in the building of modern society (Berner, 1939). However, Karl Marx and Friedrich Engels had much earlier (Marx, 1867, Engels, 1845) demonstrated that technical development in the factory will determine social relations between the owner of the technology (the capitalist) and the worker to the workers' enormous detriment. Ellul (1964) furthermore presented a pessimistic view by demonstrating the alienating effects of technology on humanity and spirituality.

The science and technology shapes society perspective is often called 'Technological determinism' which is considered by Mackenzie and Wajkman (1985) to be the 'single most influential theory of the relationship between technology and society.' It is a theory that claims changes in technology cause social change. Material forces and properties of available technologies determine social events. One criticism of this perspective is that it seems as if technology is outside of society - i.e., it can affect society without itself being affected. The questions that Mackenzie and Wajkman ask in 'The Social Shaping of Technology' (1985): What is it that shapes the technology before it has 'effects'? What role does society play in shaping technology?

2.4 SOCIETY SHAPES SCIENCE AND TECHNOLOGY

Furslang (2001) tells us that this turn around in perspective came from two pressures - the business and academic worlds. In the former category were those that believed science should be more connected to commercial purposes, and in the latter case academics sought to examine and understand the links between social drivers and science and technology. The Sociology of Scientific Knowledge 'strong program' (SSK) in the UK (Furslang, 2001) has endeavoured to study these social forces. Looking at different technologies and their alternatives at moments of choice, it is possible to see what drivers caused the decisions to be made. David Noble (1984) asks us of technological progress 'for what or for whom?' and poses that technological developments are often made in the name of patriotism, and competitiveness with the twin aims of control and domination. He proposes that there has been a shift of control to management with new technologies. We will come back to this point later in our discussions of Ursula Franklin's work.

2.5 THE INTERACTIVE VIEW OF THE SCIENCE, TECHNOLOGY, AND SOCIETY

The interactive view realizes the obvious next step that society affects technology and technology affects society. Three approaches have been identified by Pinch and Bijker (Bijker et al., 1989). The first - 'social constructivism' suggests that technology requires sociological analysis. The second puts technology in a system and integrates physical artifacts with their environment. The third approach attempts to break down the distinction between human and nonhuman actors (e.g., 'actor network theory').

A First Approach – Social Construction of Technology

In their chapter, 'The Social Construction of Facts and Artifacts' (Pinch and Bijker et al., 1989) two methodologies within the first approach are considered, EPOR (Empirical Programme of Relativism) and SCOT (Social Construction of Technology). EPOR, coming from the tradition of Sociology of Scientific Knowledge (SSK), explores three stages of scientific development. The first stage allows for the interpretative flexibility of scientific findings, the second stage brings in social mechanisms that allow scientific controversies to be terminated, and the third stage relates to the broader social and cultural context. SCOT, developed by Bijker et al., coming from sociology of technology, shows the developmental process of a technological artifact such as the bicycle, as an 'alternation of variation and selection' (page 28). A multidirectional model, Fig. 2.5(a), in contrast to the linear model used in innovation studies, considers the variants of an artifact (page 29, Bijker et al.) and allows us to consider why some die and some survive. The shaded area is filled in within the last figure shown here.

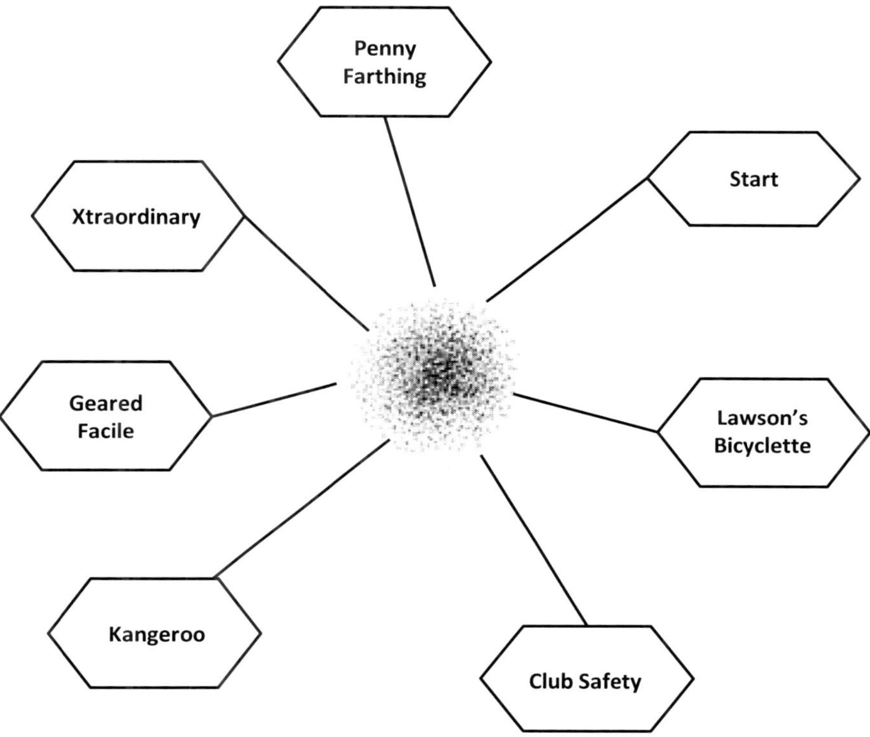

Figure 2.5. (a)

In order to see why some variants survive, we need to consider which social groups concerned with this artifact are relevant and which not – users, consumers (women, older men, young men, etc.) and in this case 'anticyclists' (see Fig. 2.5(b), page 35, Bijker et al.).

Each social group will have a particular functional need for the artifact and hence pose a certain problem, see Fig. 2.5(c), (page 36, Bijker et al.).

Each problem in turn will have several solutions, see Fig. 2.5(c), (page 36, Bijker et al.).

Finally, we can fill in the details in the original model [this is the expanded version of the shaded area shown in the first figure, see Fig. 2.5(d), (page 37, Bijker et al.)].

In this way, Bijker et al. combine the two techniques, EPOR and SCOT, to develop a fully integrated perspective.

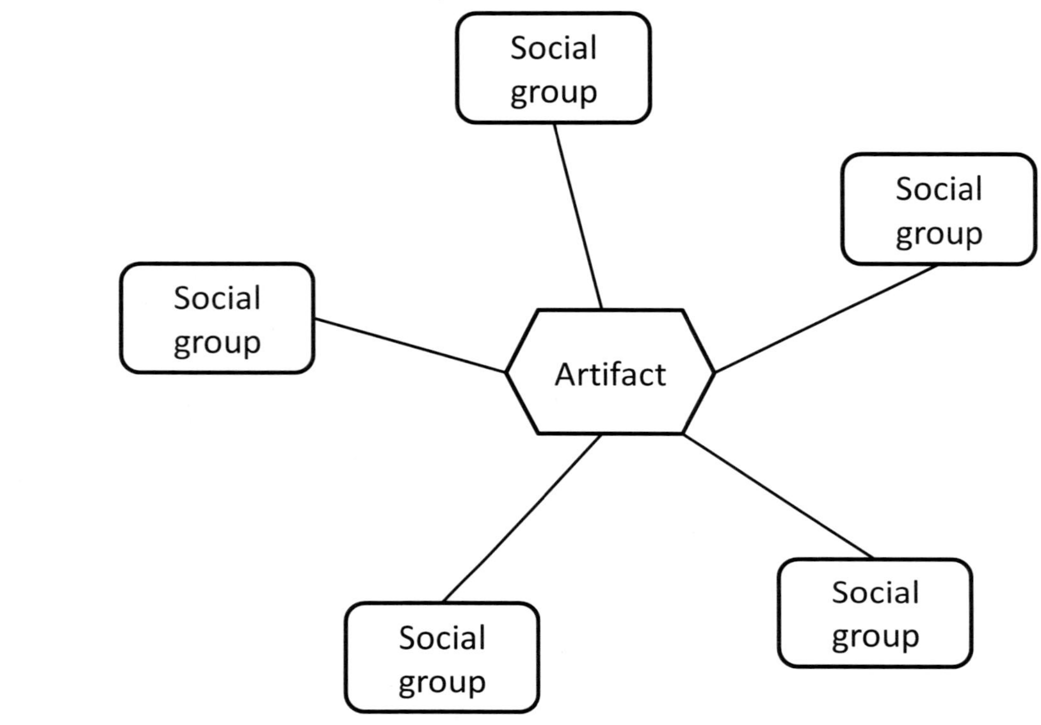

Figure 2.5. (b)

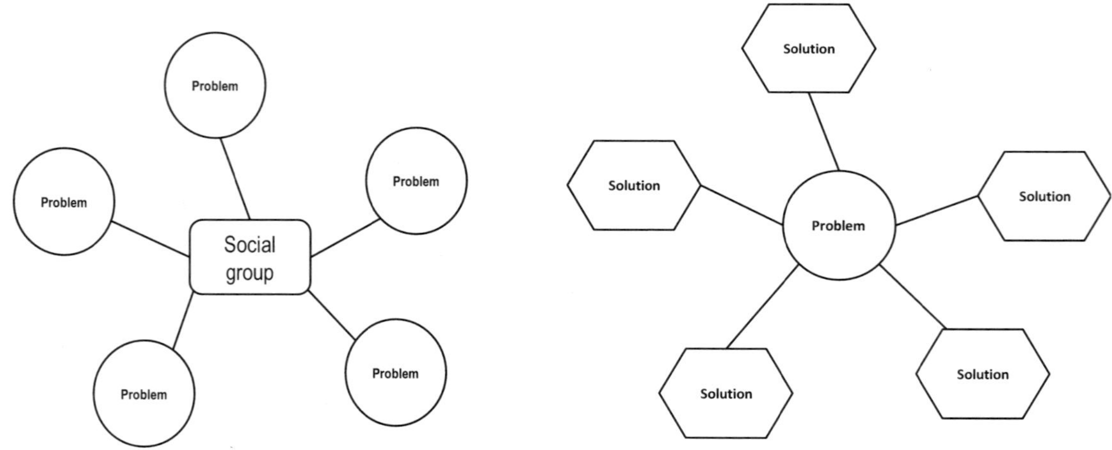

Figure 2.5. (c)

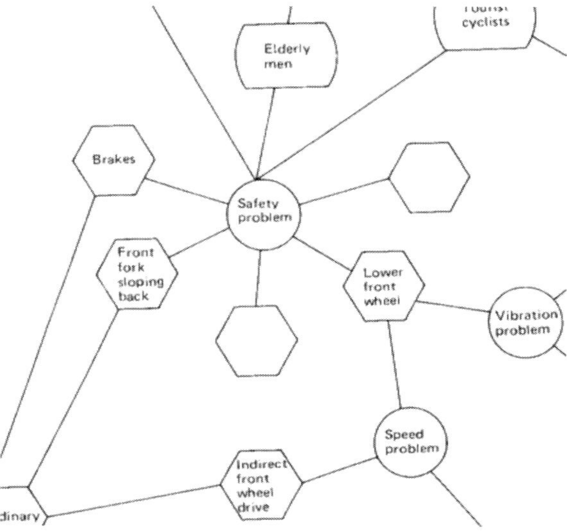

Figure 2.5. (d) Variants and development of the bicycle.

A Second Approach - Technological Systems

This is usually an area identified with the historian of technology, Thomas Hughes (page 51, Bijker et al., 1989) and considers the system of social, political, economic and environmental factors that play a role in technological development. Dependency may be in either direction. For example, 'the supply of fossil fuel is often an environmental factor on which an electric light and power system is dependant' (page 53, Hughes in Bijker et al., 1989). Recent work on technology transfer and technological forecasting would also fit into this area as well as developments using TRIZ – the Theory of Inventive problem solving which provides a series of 'tools' to identify trends in designs (see Baillie, 2004).

A Third Approach - Actor Networks

The third approach is demonstrated by the work of Michael Callon, who described what he called 'engineer-sociologists' (Bijker et al., 1989). He proposed that engineers who bring about a new technology, whether the designers, the developers or those involved at the implementation stage, actually engage in the kinds of debates that are usually considered to be within the field of sociological analysis. Callon discusses this in relation to what we would call 'forecasting' for the motorcar industry in France at the beginning of the 70's with regard to the introduction of an electric car (VEL). He considers the engineers working for EDF (Electricite de France) and the Renault engineers who came into conflict at that time.

The EDF engineers decided that they could predict where French society was going in 1973 and they innovated to create the VEL. They presented not only the technical but also the social universe in which the vehicle would function. They identified the electrochemical batteries required, a market in public transport as well as private cars, the need for cheap catalysts, as well as manufacturers that would build the VEL. Different factors rendered this vision impossible including the lack of cheap usable catalysts. Renault engineers challenged this vision by suggesting that there would be a need to set up a vast network of service stations to recharge the electrolyte and that this would compete with the oil consortiums and that they would never allow this. They also showed that protest movements against cars run on fuel were dying down.

Callon compares the two types of engineer-sociologists with the thinking of sociologists of the time - Touraine and Bourdieu, who worked on the dynamics of consumption. Touraine emphasized the role of class conflict in making society function and in producing history. He believes that technology has changed the class struggle but not in the same way that Marx proposed. On one side are corporations that control the application of technology and on the other side is the consumer who is managed by the technocrats who run the large concerns. Social movements come into play that challenge the power of technocracy. Touraine calls this class conflict 'post industrial society.' Bourdieu's society on the other hand, shows a confrontation between social agents - many different groups which are caught in a group logic organized around a dominant cultural model – that of the upper classes – other classes orient themselves around this. Whatever field it is – these classes are in constant competition – ever more so in the field of consumption.

In Touraines' model, the technologists are in control. The EDF engineer-sociologists planned the VEL project as a way of responding to French society in 1973 and to ensure they keep ahead of the game. The engineer-sociologists of Renault, however, developed a sociology not unlike Bourdieu. In Bourdieu's model, the automobile is seen as the nerve centre of society, socially embedded to the extent that it can only be transformed gradually. In the Touraine/ EDF model, the technocrats use social protest to further their aim. In Bourdieu/ Renault's case, social movements may wish to protest against the automobile but it is concluded that the only realistic strategy is to progressively introduce technical improvements, rather than replace the technology altogether.

Bourdieu's approach works in the examples given by Callon regarding the development of the motorcar in France in 1973, but the author stresses that the explanation of the preferences of the consumer omits many of the elements which make up these preferences. It is by chance that in this case Bourdieu happens to be right and Touraine wrong. If there had been a cheap catalyst which worked as a substitute for platinum, it could have been the other way around.

Callon, therefore, introduces the actor network theory (also associated with Bruno Latour who developed it further) as the breaking down of distinctions between human actors and natural phenomena. This approach reverses the usual relationship between participant and analyst and casts the engineers as sociologists. An actor network is 'simultaneously an actor, whose activity is networking heterogeneous elements and a network that is able to redefine and transform what it is made of' (page 93, Bijker et al., 1989). Most importantly, it networks such items as fuel cells, electric cars, electrons, catalysts, industrial firms and consumers, e.g., it includes heterogeneous elements, both animate and inanimate, that are linked for a certain period of time.

Sismundo (2004) tells us that it may be considered a materialist theory as it explains the successes and failures of facts and artefacts; they are the effects of the successful application of actions, forces and of human interests. It can also be considered a relationalist theory with innovative results. He believes it is not without its problems, but it 'stands as the most successful of S&TS's theoretical achievements so far' (page 74). Some of the problems he identifies are as follows:

1. It does not pay attention to subjective factors such as culture

2. The analysis tends to centre around key figures and may encourage the following of heroes and anti heroes thereby missing the real story

3. It resists the idea of intuition that scientists discover rather than help create properties of natural things

4. Science is considered to be objective and any multiple interpretation of data is neglected

> **Challenge Box:** Think of a common product of engineering that you encounter everyday. Try to imagine what the factors would be within an actor network for that product.

2.6   TECHNOLOGICAL PRACTICES

Ursula Franklin is a key thinker in the area of engineering and society. She has given a series of important lectures published under the title 'The Real World of Technology' (1990). She considers technology to be defined as 'practice' or 'ways of doing something,' which links to culture as culture is a 'set of socially accepted practices and values.' She suggests that:

'A way of doing something' is 'holistic,' she says, when the doer is in control of the work process. The way of doing something can be 'prescriptive' when the work is controlled and divided into specified steps each carried out by an individual.

Holistic Versus Prescriptive Technologies

Holistic technologies, Franklin tells us are usually associated with the notion of craft – the doer is in control of the whole process. Prescriptive technologies, on the other hand, are based on division of labour. Making something requires it to be broken into steps, each step carried out by a separate worker or group of workers. We mostly associate the beginning of this kind of productivity with the Industrial Revolution although Franklin reminds us that it began much earlier, before 1200 BC with Chinese bronze casting. Franklin, herself a metallurgist, has studied this process in detail, and it was this that first caused her to fully understand prescriptive technology. The control over the work moves to the organiser or the boss. She describes it as designs for compliance – everything is prescribed with precision. External control and internal compliance are seen as necessary.

A Culture of Compliance

Franklin explains that as much as this culture within technology has created products in numbers and qualities, that was the intention; this acculturation to compliance and conformity has accelerated the use of prescribed technologies in administration. Government and social services, she points out, have 'diminished resistance to the programming of people' (page 25). Bureaucracy then, is the 'acculturation into a culture of compliance built on the willing adherence to prescription and the acceptance as normal of external control and management' (page 116). We get used to certain ways of behaving, and it becomes 'common sense' to us.

> **Challenge Box:** What do you think about Ursula Franklin's idea that we are somehow the subject of a 'culture of compliance'? Can technology affect our way of life in this way?

Reciprocity and Indivisible Benefits

A further important issue elucidated by Franklin concerns the separation of expertise from direct experience and a need to evaluate experiences of those at the receiving end of the technology. For example, she tells us that 'Communications' technologies have become 'non-communication technologies' - the absence of reciprocity is developed. We cannot feed back to the television when we don't agree – as we could in a live talk or debate. She reminds us of a Ben Wicks cartoon showing a repairman in a living room removing a television set with a smashed screen. A man on crutches with one foot heavily bandaged stands next to the set, to whom the repairman says, 'Next time Trudeau speaks, just turn the set off.'

Indivisible Benefits

One area which is always considered 'public' engineering, is the design and building of roads and airports, paid for by the public. Franklin explains the difference between divisible and indivisible costs and divisible and indivisible benefits, by giving the example of growing tomatoes. You have a garden and your friends help you to grow a great tomato crop and you can share this with those who helped you. This is a divisible benefit. If you work to fight pollution in some way, those who help you get benefits, but those who did not help you also get benefits. These are indivisible.

Since the Industrial Revolution, publicly financed infrastructures were created supporting new technologies. The public provides the space permission and finances for much of the research, regardless of who owns the railways or transmission lines. However, Franklin fears that governments are increasingly neglecting many things we hold in common or in the 'common good,' as 'indivisible benefits.' She describes how infrastructures that are publicly funded have become 'divisible benefits' to the private sector. Indivisible benefits, on the other hand, such as clean air and uncontaminated water are less and less safe guarded. Furthermore, she points out that the needs to provide for certain technologies might require training in computer literacy at the expense of other areas such as history or moral literacy. She asks us: Which education is more in the public interest?'

Challenge Box: Can you think of one indivisible benefit created by some project that you or someone you know did?

Choices

Franklin goes on to explain that values of technology have 'so permeated the public mind that all too frequently what is efficient is seen as the right thing to do'(page 123). She would like to see public discourse break away from this mindset to one which 'focuses on justice, fairness and equality in the global sense.' She has made various choices in her own career, such as not to work on projects related to atomic energy because she finds it 'unforgiving' and 'unforgivable.' She would work on nuclear waste disposal but only after Canada has agreed to discontinue building nuclear reactors.

We often believe we don't have choices, but as she clearly points out, this is not the case. She asks us, when deciding upon a particular project, not to simply consider benefits and costs but to ask 'whose benefits and whose costs?' (page 124).

Franklin makes a series of recommendations to do this – to ask the questions or a particular public project. Does it:

1) promote justice

2) restore reciprocity

3) confer divisible or indivisible benefits

4) favour people over machines

5) minimize or maximize disaster

6) promote conservation over waste

7) favour reversible over irreversible

Franklin's final gift to us in her series of lectures is the idea of bookkeeping. She believes that we need three sets of books: one for the economy, one for people and social impacts, and one for environmental accounting.

> **Challenge Box:** Imagine you are to create an account book for one recent technological development. Consider the aspects that need to be accounted for in relation to the social, environmental and economic.

Preventative Engineering

Franklins' work has been continued by Bill Vanderburg (1985, 2000) who has also clearly been influenced by his teacher Ellul (1967) where he studies the interconnections with other things within a wider whole. His main contribution is the notion of 'preventative engineering' where he looks at cost – balance, e.g., how expensive it is to have health care as one of the major expenses of a company – if we look after the worker we will pay out less. He claims that it is possible for companies to make a profit by taking care of the environment and social impact – and that in the long run this will be a financial gain for them. If companies plan for impact, then they will 'prevent' the costs of health problems, liability and environmental implications.

2.7   FINAL THOUGHTS

In this chapter, we've taken a look at some of the most recent work on the relationship between science, technology, engineering and society. Social and environmental problems could be solved by technical means. They could also be caused by technical means. We have summarized some of the warnings of those that care, together with some of the ideas and tools for moving forward into the future, with respect for the environment and for all people in local and global societies. There are complex maps which relate the various actors and artifacts together with the social, political and economic environment in which they work and act. In such a short space we can only hope to capture the essence of this complexity and capture the real sense of responsible engineering - which is connected to the people whose needs it serves.

CHAPTER 3

Engineering and the Public

Figure 3.1. "Now, why would we purposely build the bridge too low to let Poorburb buses into Richville?"

3.1 INTRODUCTION

In the previous section, we dealt with the relationship between engineering and society as a whole – how it might shape or be shaped by the way that we live in any particular context or time. In this section, we are going to look closer at the relationship between people and science, technology and engineering. Again, most of the work in this area discusses 'science' but means all three. There has been much recent debate about how to get the public more interested in science, and how to get scientists more interested in liaising with the public; however, it is not as simple as finding ways of promoting this conversation. The government and industry in the UK, in particular, has become very concerned with recent events such as BSE and foot and mouth disease, realizing that the British public will no longer simply accept what the press tells them. They appear to be much more doubtful of scientists' findings than citizens of the US and Canada. We are going to explore why this might be and whether the British public has a right to be skeptical. Because of the controversies, however, it is true to say that there is far more interest in this topic in the UK currently, and so we will draw on many British reports from science and government departments and associations, as well as literature which deals more broadly with the questions raised.

3.2   UNDERSTANDING DIALOGUE

Figure 3.2. Dialogue or lecture?

Twenty years ago, it was commonplace to hear discussions about the 'ignorant public' and discussions ranged around the topic of 'Public Understanding of Science' (PUS) – transferring

Figure 3.3. What is dialogue?

science to the public who knew nothing and didn't much care to. More recently, however, in the UK, it is even considered a bit patronising to use the expression 'public understanding' or even science communication although these expressions are still found in other English speaking parts of the world. The UK government is even setting up 'an Expert resource Centre for Public Dialogue on

Science and Innovation to assist all parts of the government in enabling public debate on science and technology related topics (HM treasury, 2006).

The UK Royal Society set up its Science in Society programme in 2000 (UK). It is stated (Bhattachary et al., 2004) that 'the programme embodies a core value of the society – that of furthering the role of responsible and responsive science, engineering and technology in society' (page 4), and it clearly states that dialogue is the key approach. In 2001, the dialogue initiative was created in which more than 600 people from across the UK were involved in discussions about science and science policy, e.g., 'Dialogue on Genetic testing.' The discussions were then included in the white paper 'Our inheritance our Future.' Other schemes include the MP pairing scheme, which forms links between 42 members of parliament and research scientists. The 'Genetic Futures' project has involved 800 school children- 14-16 years old to discuss science and ethics.

But what do they mean by dialogue? Usually, this is where the communication is two way and 'negotiated,' i.e., it is not simply the case of one person telling facts to another. Ideally, the information is shared and discussed, and everyone may learn something new at the end of the conversation.

Most UK government reports offer a range of reasons why dialogue is necessary:

Increasing democracy by promoting open and transparent decision making, greater trust and confidence in the regulation of science and the decisions taken (Whitmarsh et al., 2005, page 9).

Supporting democracy and making better decisions (POST, 2001).

Meeting the productivity challenge (HM Treasury, 2006). 'The Government's goal is for the UK public to be confident about the governance, regulation and use of science and technology – by both government and business – and to be actively engaged in scientific debate' (page 58).

Building confidence in decision making related to the undertaking, development and overall governance of science and technology (Kass, 2006, page 1)

We also see more broadly defined reasons for dialogue, e.g., 'dialogue is part of good citizenship' (Whitmarsh et al., 2005, page 18, Irwin, 1995, Leach et al., 2005).

Challenge Box: When was the last time you heard the issue of public scientific knowledge being discussed? Did they promote dialogue or were they more focused on public ignorance?

However, it's all very well stating that dialogue is important, but we have to take care that this is not lip service so that the public will be quiet and not make a fuss, which could affect political leaders or corporate money making.

The above information focuses on two issues, *democracy* and *confidence in decision-making*. Let us look at these two in more detail.

Democracy – What exactly are the UK Government reports referring to when they consider supporting democracy? What can the public actually influence? The British Association of Science (BA) 2005 report (Whitmarsh et al., 2005) states up front that 'it is not the purpose of dialogue to intrude upon discussion amongst the science community about scientific knowledge. There is no suggestion that the progress of scientific ideas should be democratically decided.' This appears to be speaking to the scientists, many of whom they anticipate would be horrified that their scientific agenda might be put to public vote. On the other hand, much has been written about 'upstream dialogue' – getting the public to intervene in the decision making process at an early stage. Sir Aaron Klug, president of the UK Royal Society, seems to be quite clear about his aim that scientists need 'input from non experts to make us sure that we are aware of the boundaries of our licence.' So, a sort of check against our own consciences?

Confidence in decision making – Is this primarily what the government is interested in? – that the public trust them to make decisions and do not interfere with their fears and concerns that the public does not probe. The BA report tells us (Whitmarsh et al., page 20) 'what we do not know is on what issues and for what reasons, the 'average' person would consider taking action…there is little data on the general question 'what would you march for?'

In some countries, it is much more likely to see a march or protest than in others. In the 1960s, people were protesting all over the world against the Vietnam war. Although it is said that the protest was against war, it is clear than contemporary wars do not create the same level of angst. Is this because Vietnam had general conscription or are we less likely to march today in France, England and the US? In Argentina and in many countries in South America, things are very different. People in Buenos Aires are protesting every day about something they value. 'Recovered factories' refers to a movement in Buenos Aires, Argentina following the economic crisis of 2000/2001, whereby workers in factories where the owners have left and declared bankruptcy, decided to open up the factory again by themselves and gain legal status for doing so. They wanted to work and support one another to try to open the factory. Although many have closed since then, many are still operating in 2008 (See Naomi Klein and Avi Lewis's documentary 'The Take' for more information). These factories are examples of workers taking their own decisions to act. Their need was to work, a factory lay unused and they had the skills to make it happen. However, the fights are still continuing and during 2007 workers at Hotel Bauen, which had been operating for several years under their control, were evicted. The protests continue.

Figure 3.4. Recovered Hotel Bauen workers in Buenos Aires protesting against their eviction.

Challenge Box: What would you march for?

In a 2002 report by the Parliamentary Office of Science and Technology (POST, 2002), UK, it seems clear that some good intentions about policy were intended:

(The review board for radioactive waste) 'will need to remember one simple principle of public engagement: it is a waste of everyone's time unless the decision-maker is willing to listen to other's views and then do something which it would not have done otherwise' (page 2).

The POST was concerned that dialogue not be for the purpose of (page 3):

Avoiding decision-making for as long as possible

Suppressing expression of disagreement in an overcompelling search for consensus, thus pacifying potential opposition rather than building agreement

Simply pandering to public concerns without taking due account of the diversity of views alongside scientific and political factors

Deflecting blame from those responsible and accountable for decisions onto participating citizens

Despite the POST views to the contrary, most reports do seem to infer that democracy and confidence in decision making refer to the public being educated and aware enough of the science in question so that their (supposedly unfounded) fears may be suppressed and the science may be developed, allowing for development to the market place, where it can do its job of enhancing the economy (HM treasury, 2006). Democracy in this case is limited to decisions about whether or not a technology should be allowed to be implemented once created. Upstream dialogue suggests that the earlier this happens the better, which might be related to the fact that if we stop some research at the last minute, it will have wasted a lot of money but is also likely to be connected to the 'ownership' of ideas. If the public have been consulted early enough, they will feel safe sooner and even be transformed into advocates for the technology in question. It could be about making sure that the public have been consulted so that they don't actually stop the process – or slow down too much the 'technology transfer' (laboratory to market). Why is it, for instance, that genetically modified foods are so unwelcome in the UK? The British public seem to have actually put a halt on the import of certain agricultural products. One view is that were they consulted earlier this block would not have happened.

However, faulted and problematic the process, there is an attempt in some places to actually consult with the public, for whatever reason and to take into consideration what they say. However, as we have seen, this is limited to making decisions about an already developed or partially developed technology. The BA report suggested that we must not intrude upon discussions in the scientific community, nor may the progress of scientific ideas be halted. This, in my opinion, is where the problem lies. How far upstream is the government prepared to go?

> **Challenge Box:** Find one recent article where there is a controversy between the public and government or industry about a technological issue. Was there any consultation before the technology was created or was the controversy only after the fact?

3.3   NEEDS ASSESSMENT AND LOCAL KNOWLEDGE

We need to consider better ways of negotiating with the public to assess needs *before* a new technology is created. It is often stated that the 'lay' public don't know enough science to be able to contribute to the discussion (Sclove, 1995). However, it is certainly possible for them to comment on their needs. In Fig. 3.5, my student Thimothy is asking a local villager in Roma, Lesotho, if they would like to think about what the material he is holding in his hands which is made from waste plastic and local plant fibre. Is this something he could think of a use for? What plants do they grow there? This kind of exploration is in contrast to market research, which is where a corporation tries to find out whether or not the public would buy a specific product – whether they 'want' the product. There are several groups that actively try to create programmes for community groups and citizens to be responsive to needs analysis (Sclove, 1995). Furthermore, there are case studies which show that some of the most effective and democratic political action is driven in large part by responses to very real personal and economic need. In his book, 'Street Democracy' Barker (1999) 'inquires into how people create and use local political settings or collective public actions located in space and time close to where they live and work..' (page 21). Again we are reminded of the recovered factory movement in Argentina.

   A good example of this is described by James Coburn in his book Street Science (Coburn, 2005). Citizens in New York became worried about air quality following 9/11. Locals could tell through their experience that the air quality couldn't be good enough. 'How can two 100 story buildings that were never meant to disappear into thin air – given all the things in those buildings that were never meant to disappear into thin air – and that air be perfectly safe to breathe? (page 1). Coburn follows one community and how residents organized community knowledge to improve scientific enquiry. He tries to revalue different forms of knowledge. When locals experience conflicts with the conclusions of experts, questions arise about how professionals 'create define and prioritise problems which warrant attention' (page 3). He remarks that the World Development report (World

Figure 3.5. Assessing local needs in Lesotho.

Bank 1999) notes that 'local knowledge, on par with additional capital, is the key to sustainable social and economic development, reducing poverty and improving health.' However, he further comments on how this doesn't seem to be taken into consideration on home territory in the US urban settings where populations are largely the poor, immigrants and people of colour. Coburn 'raises the contentious political questions that professional 'techno' science tries to silence by often claiming that an issue is 'purely technical.' He builds on a number of existing participatory models of knowing and doing (participatory-action research and community-based participatory research)(PAR). PAR, he says, makes action research more democratic - those impacted by the action must be involved in practice, and it makes the methods of enquiry more legitimate by opening up to lay members. He worries about consensus building. He says that problem definitions, meanings and purposes are not usually open to negotiation but should be. Guidelines for public participation are available (Creighton, 2005) built on expertise based on experience with public meetings in many discipline

Figure 3.6. Chilavert 'recovered' printing factory.

areas, group dynamics including interpersonal communication skills, design of interactive meetings, values research, communications theory, political theorists and participatory management literature, deliberative democracy and risk communication. A guide to becoming a researcher for social change is also available (Morris and Muzychka, 2002).

One method that Coburn presents is mapping of local knowledge. He presents the case of community generated maps of local toxic sites in Brooklyn, New York by a group called the Toxic Avengers. In order to create his case study, he used interviews, observed meetings, shadowed activists and community researchers and did 'street' ethnography - unstructured interviews and conversations with community residents about issues and processes discussed in the book.

A group of high school students in the Brooklyn Greenpoint/Williamsburg neighbourhood created a map, annotated with toxic hazards. The school students were in a science class, and they were doing a unit on understanding the neighbourhood environment. The class instructors organised

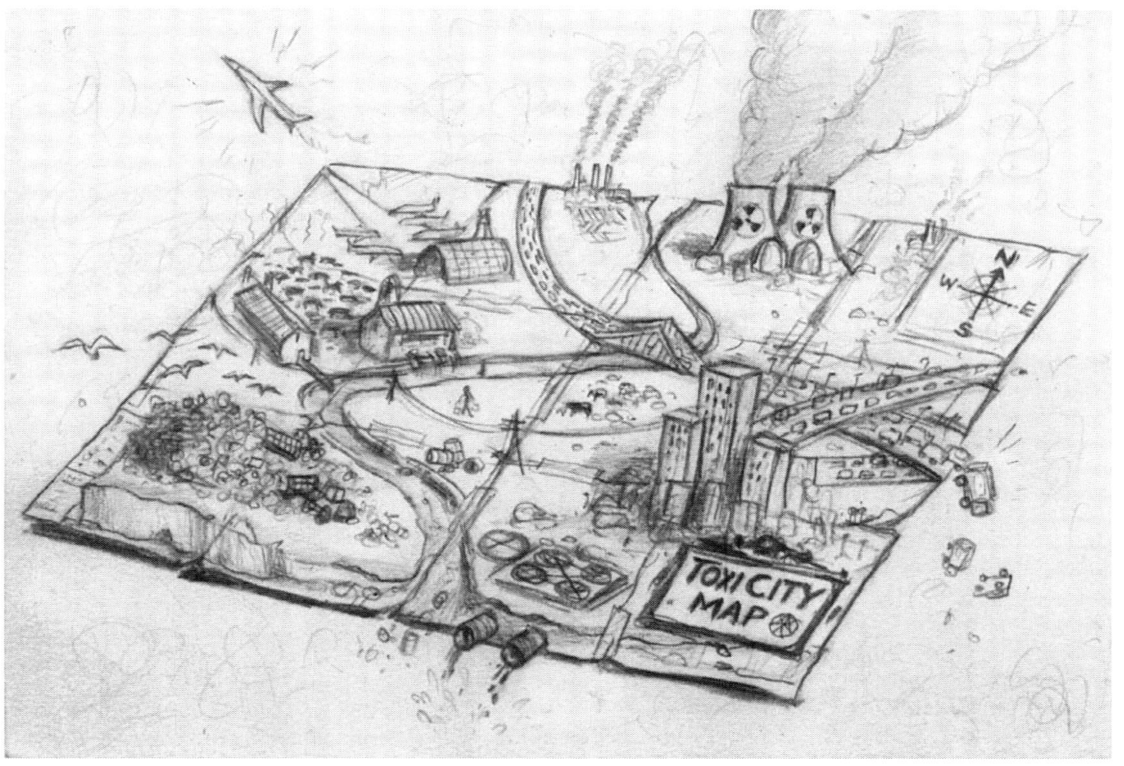

Figure 3.7. Creating a map of local knowledge.

a tour and then asked students to create a community risk map – listing hazards, code and symbolize the hazards and map on a poster board. The student group called their map 'Our Town' and became very motivated. They took the map to Luis Garden-Acosta, El Puenta's (community group) founder and executive director, to encourage him to invite the Hasidic Jews to a joint meeting.

'It was nothing but confrontation (with the Hasidim) before young people from the Toxic Avengers came to me and said 'Isn't it time to ask the Hasidim to join forces with us in reclaiming our environment?' It was their request and their graphic map gave me the 'ah ha' that we all breathe the same air' …It was a historic moment when the Hasidic Rabbi, a leader of the Satmar, walked in through the doors of El Puente. We were planning a march to a hearing on the Radiac emergency procedures and Rabbi Niederman volunteered to help lead the march through Latino streets. I can't describe what a change that meant' (Hevesi, 1994).

The Toxic Avengers' maps did not directly influence professionals, but by helping organize the community around environmental issues, the student maps helped build a coalition that played a role in influencing professional decisions - the 'Community Alliance for the Environment' (CAFÉ) - the first multiethnic environmental coalition.

Challenge Box: Consider building a local knowledge map in your area. What could it be used for? How could it help? Who would benefit?

Coburn believes that maps perform three political functions:

- Aggregate and select data

- Create identity forming devices

- Make boundaries

He firmly believes that local knowledge can point out low-cost and more efficient intervention options. Furthermore, he says that local knowledge improves environmental health research and policy making in the following ways:

i Aggregation –professional decision making tools always aggregate and miss particularity
ii Heterogeneity – local knowledge can highlight how professional models pay inadequate heed to the inter-individual or intergroup variability of the population
iii Lifestyle – causal factors – can label some things as not relevant. This does not therefore express the holistic way in which someone moves through the world
iv tacit knowledge – unspoken information

Coburn introduces us to two terms:

> Distributive justice – local knowledge can raise distributive justice concerns in disadvantaged communities.
> Distributive justice is the right to the same distribution of goods and opportunities as anyone has or is given.

The World Social Forum has become the place where local knowledge is shared among activists concerned with scaling up what they know into a global social justice agenda.

Figure 3.8. Embracing the world social forum.

Gupta (1999) tell us 'formal scientists do not recognize, respect and reciprocate the informal scientific knowledge, creativity and innovation at grassroots level in society. The science underlying the successful overcoming of some of the day to day struggles of economically poor but knowledge rich people does not get articulated or acknowledged..' (page 368). He worries about a tension between standardized knowledge and diversified need – 'for a large number of people living in the high risk environments such as drought or flood prone regions, forest fringe areas, mountain areas, etc., there is not much scientific knowledge available that can improve their livelihoods…. organisational incentives for generating technologies with limited potential for diffusion are very low' (page 368).

His solution to move beyond the barriers blending excellence in formal and informal sectors is the 'Honey Bee Network' in which 75 countries draw on technological and institutional innovations for sustainable natural resource management developed by people unaided by NGOs, market or state.

Figure 3.9. The Honey Bee Network.

He suggests the following to help bring informal knowledge to the formal sector and for scientific discussions to be more accessible (Fig. 3.10).

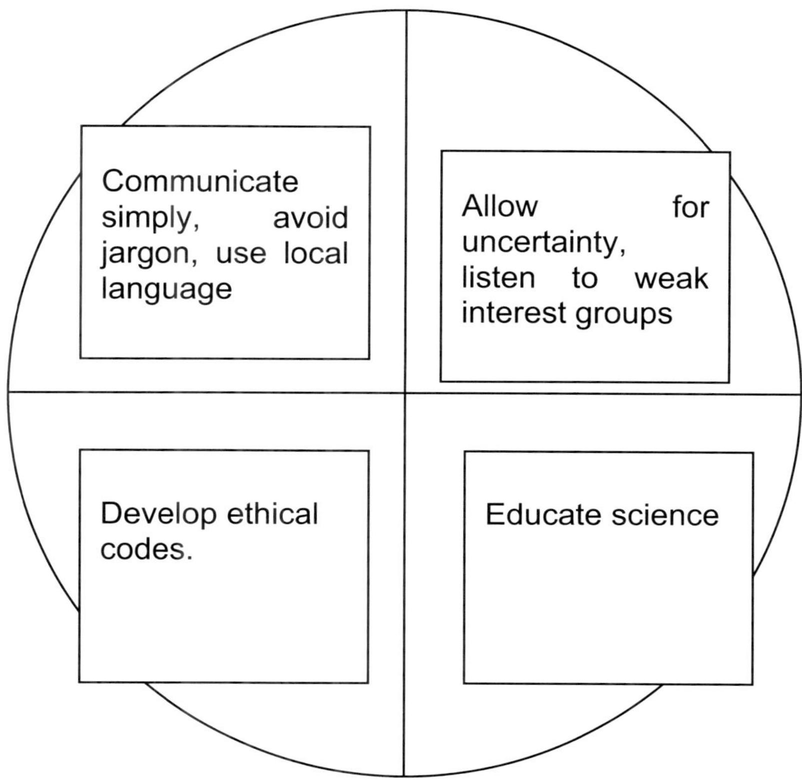

Figure 3.10. Informal meets formal knowledge.

Gupta makes some interesting points relating to ethics and education. He suggests that the place for ethical code generation should be the professional associations but instead of concerning themselves with this important role the associations have focused on the career profiles of a few professionals. There are not many such associations anyway in developing countries, and when there are, they tend to insulate scientists from their social base rather than promote the dialogue. With regard to education, he notes that textbooks make hardly any reference to people's ability to solve local problems creatively and in an innovative manner. The blending of local knowledge is ignored.

> **Challenge Box:** Look up the Code of Engineering Ethics for your discipline/ professional society? What do you think of the above statements by Gupta in the light of your code?

3.4   CITIZEN SCIENCE

Irwin was probably the first to use the expression Citizen Science (Irwin, 1995) when he addressed the first public debate on environmental issues which arose in the 80's. He questioned the notion of the expert and whose knowledge was important (Fig. 3.11).

Citizens became their own resource. He thought that we should integrate science with other assessments to create our problem definitions. He also promoted the integration of social technical and environmental lenses to view problems. Since then this way of thinking has developed into sustainability studies. In the area of Science and Technology Studies and Development Studies authors have been thinking about Indigenous knowledge for some time. Leach et al. (2005) tell us that the first wave of science studies was to understand, explain and reinforce the success of science without questioning the basis, so the perceived crises of legitimacy among publics were deemed to be because of pubic misunderstanding. The second was to challenge assumptions and practices so science was reconceptualised as a social and political activity. This therefore challenged the distinction between expert and non expert knowledge but didn't go further to look at non expert knowledge. Citizen science comes from this. The third wave which concerns public knowledge or experience is much more recent (Collins and Evans, 2002). The term cognitive justice is used to describe different forms of knowledge and associated practices and ways of being.

Sci-Fi author William Gibson tells us that 'the future is here. It's just not evenly distributed yet.' He is quoted by Diane Warburton and Jack Stilgoe for the Science Horizons project (Stilgoe and Warburton, 2006), who are concerned that 'until now the future has not been very democratic.. The future is seen as a place that only a few people have access to, because they are experts, or they are visionaries, or they are people powerful enough to make it so…The 'great man' view of the future is driven by technology…As well as being an unreliable vision so the futures that are driven by technologists normally exclude any sort of discussion of what people might want…It is not coincidence that many pictures of the future leave out people. Monorails and hovercars fly

Figure 3.11. Who is the expert?

facelessly beneath cloudless skies…the utopia that Gates paints is a geek's utopia, focusing on a closer relationship between the individual and his computer. This is far from everyone's ideal world…the question is: How can we put people back into the future?' (Stilgoe and Warburton, 2006, page 4).

This, they say is the first paradox of engagement (page 6). 'We need public engagement to take place when decisions are still open, but this means that it might not be always exactly clear what we are talking about. . we need to think about how we start public conversations about science and technology in general.' The second paradox (page 7) is that we need wide debate with a wider cross section of society. Increasingly, activities are promoted which are very popular especially across Europe but increasingly also in the rest of the English speaking world including science centres, science cafes where discussions are hosted, science shops where scientists help to solve public issues, museums and festivals.

The Science Horizons project, UK, (Stilgoe and Warburton, 2006) had a year of activity during 2007. They have created a series of interactive scenarios on the computer about the future of science and technology in people's lives and have been running workshops with different groups of people to collect opinions.

The Science Horizons programme declares itself to be the first public engagement exercise in the UK to focus on the potential future uses for science and technology, using a set of fictitious potential scenarios set from 2025.

Challenge Box: Go online at sciencehorizons.org.uk. If it is not still active at the time of reading, locate another science/engineering-public interaction site on the web. Try to take a position about the website and its intentions and compare it with the authors of the site.

3.5   RISK AND UNCERTAINTY

One of the main reasons given for public dialogue with science is that we need to help the public understand how to assess risk and deal with uncertainty. It is stated, for example, that 'risk and ethics are often conflated; it is not always clear whether objections are on moral grounds or on grounds of possible consequences or both..' (Whitmarsh et al., 2005, page 10). The suggestion is that when the public make complaints about the risk of something such as a pesticide or nuclear waste, that they do so on the basis of the fact that it is 'wrong,' but in fact, they are more concerned about their own safety. What is 'wrong,' which may be related to moral reasoning, is likely to be connected to a person's values – what we value in the world. If I value beautiful, smooth, green lawns in my garden, over the consequences of pesticides on the environment and on other plants, insect and animals as well as the effects on health of human beings, then I would support the use of pesticides (you see where the author's bias lies!). Sometimes it is difficult to work out why we think something is wrong – it may be because we have been told this so many times by our parents or the media.

It has been suggested (Whitmarsh et al., 2005) that the British public think scientists will provide 'facts' but distrust them to communicate the ethical and social implications. This is interesting as it suggests that the 'facts' are solid, unmovable. It is always suggested that the public cannot deal with uncertainty – that this will make them very afraid. So scientists are encouraged to provide 'facts' and the government persuades the media to make stronger facts out of press releases from researchers. A story about research towards a possible cure for cancer gains the headline 'the cure for cancer has been found.' However, this can backfire tremendously, and the British public have recently stopped trusting the newspapers as much since they have been told too many facts which were then disproved – in critical areas such as foot and mouth disease. In fact, most people want to know the balance of probabilities – the range of possible data so they can make up their own mind. They will anyway, so why not provide more data for them to do so.

Gupta (1999) tells us that dealing with risk and uncertainty has a lot to do with control versus accountability. When policy makers want to reduce their zone of responsibility, they complain about uncertainty. Policy makers often want the answer in a hurry. Gupta also believes that scientists perceive risk differently – they ignore highly familiar risk such as speeding or smoking, also new risks attract attention so AIDS is discussed more than chronic risks such as TB and firearms. Furthermore, familiar problems with a new uncertain cause – e.g., breast cancer, gets more attention that the known causes.

A classic case from 1980 demonstrates the dilemma. The National Union of Agricultural and Allied Workers (NUAAW) were engaged in a dispute with GB regulatory agencies. Herbicide 2,4,5-T was considered to be allegedly hazardous causing birth defects and cancer. The farmworkers had issued a questionnaire to many local farmers and asked them 'do you suffer from the following? They created a series of 14 horrible case studies. The ACP (Advisory Committee on Pesticides) responded this was not causal – i.e., that they could not prove the cause of the problems. Farmers, however, suggested that a balance of probabilities should be enough, and where lives are at stake, we cannot wait for certainty (Irwin, 1995).

Figure 3.12. Making difficult decisions (see vol. 2).

We can see that although the public may not like uncertainty – it is, moreover, the multinationals like Big Pharma that try to squash uncertainty. They do not like us to know that there is any doubt about the safety of a product. We, the lay public, might well choose to abandon their product, not for unnecessary fear, but through good sound logic that its not worth the risk. There have been countless stories of risk being suppressed.

My first job out of university was in public relations. I wanted to communicate with people and not be stuck in a laboratory. I worked for a small PR consultancy and was asked to contribute news stories on technical subjects to different engineering magazines. Sounds innocent enough. However, I had only been working there for a few months when I was asked to write an article about asbestos for a British company who still produced it. I was to discuss all the benefits of the fibre. It is, indeed, an incredibly strong natural fibre, which was why it was used so much in ceilings and other building materials.

However, during the course of my work, I saw the documentary by John Willis, originally made for Yorkshire Television and first shown in July 1982. This documentary 'Alice, a Fight for Life' showed the life of Alice Jefferson who only worked at an asbestos factory for three months when she was 17, and 30 years later she was diagnosed with mesothelioma. Legal battles are still going on (Asbestos Hazards Handbook, 1995), but the T&N company was accused of knewing about the asbestos and covered this up for many years so that they could carry on making profit. They may have convinced themselves that there was not enough evidence and that the public should not worry about the uncertainty. They are paying for it now (or the company that has taken them over), but it serves as a warning about the profit motive, risk and uncertainty.

On December 12, 1996 the Broadcasting Complaints Commission (BCC) found against T&N in a complaint arising from a BBC television documentary entitled: An Acceptable Level of Death, which was broadcast in 1994. T&N claimed that the BBC's presentation was unbalanced and misleading: unrepresentative documents on asbestos and health had been selected, the company had been unfairly accused of failing to act on health warnings given in 1958 by Dr. C. Wagner, evidence presented by a former Medical Officer for a T&N subsidiary was compromised.

The Commission in the case shown above found that the statement that "T &N had known more about the health risks from asbestos than they had been willing to disclose to their workers or the outside world" had not been unfair. The fact that 'important research by Dr. C. Wagner (1958) was "apparently" not sent to T&N's health committee satisfied the Commission that the BBC's assertion that the company "had not gone far enough" in protecting their workers from health risks was fair..' (British Asbestos Newsletter, Issue 26, Winter 1996/7.)

There have been many further reports since this time.

ASBESTOS VICTIMS' COMPENSATION BATTLE
Edmond Bracewell
EDMUND BRACEWELL | Battles on for life and compensation.
Asbestos is Britain's biggest industrial killer. Hundreds of people in the North West alone are dying from asbestos-related cancer. Yet the hazards of the material have been known about for over 100 years, according to Geoff Tweedale from the Manchester Business School. Geoff says, "The Government knew as early as 1900. The factory inspectors - the employers - must have known at roughly the same time but certainly by 1930."
Inside Out - North West: Monday 20th January, 2003
bbc.co.uk

In the article below, you will see that, in fact, public relations is, indeed, blamed for shedding a positive light on the company. It is hard to believe that I was once in a position to write such an article. I resigned from that job without writing the piece. It gave me courage to do the same thing again – to make decisions based on what I value. Resigning cost me nothing – I would get another job. It is not always so easy for people to resign. In Volume 2, we explore ethics and personal values, more and explore issues of personal choice and what this means in more detail.

Interestingly, the British Association of Science report 'Connecting Science: A Review of Recent Literature on Science and Society' (Whitmarsh et al., 2005) has pointed out that closer links between industry and universities causes public distrust in scientists in the UK. There is a general understanding, at least in the UK, that if an industry has funded research then it will not publish the results publicly unless they are positive.

During the school holidays, Robert Ballantine played with his friends on the riverbanks near his Sunderland home. During the dry Summer days, there was a lot of dust but the boys overlooked this. They weren't to know that the dust contained asbestos and that this dump, larger than the Royal Albert Hall, belonged to a subsidiary of the UK's biggest asbestos company. Mr. Ballantine died at age 50 from malignant mesothelioma on July 13, 1998. ... speculation has continued over the company's state of knowledge, effect on government health and safety legislation, levels of workmen's compensation, legal strategies and attitude towards dust control. These and other subjects are dealt with in remorseless detail by Dr. Geoffrey Tweedale in a new book: Magic Mineral to Killer Dust, Turner & Newall and the Asbestos Hazard (Oxford University Press 2000). ... A thorough, and at times, harrowing examination of medical files, company reports, memoranda and correspondence led Tweedale to conclude: "The company's attitude towards matters of health over so many years may be regarded as strikingly irresponsible. In the last decade or so, T&N has tried to defend itself in court actions by arguing that it has always applied government safety regulations, that it has always adequately warned workers about the risks, that it has paid 'fair' compensation, and that it has supported medical research. Its archive shows such claims owe more to public relations than to fact. .it neglected to implement ..schemes fully in both the UK and especially overseas; it failed to warn customers; refused frequently to admit financial and moral liability for the consequences of its actions; often paid only token amounts of money for industrial injuries and deaths; tried to browbeat doctors, coroners, and the Medical Board; sought to suppress research linking asbestos and cancer; gave the government inaccurate data about disease amongst its shipyard workers; and disseminated imprecise information about the 'safety' of asbestos."

British Asbestos Newsletter Issue 37 Winter 1999/2000, compiled by Laurie Kazan-Allent.

Challenge Box: What do you think about asbestos still being sold? Do you know which other materials are carcinogenic? Should they be made and sold? Under which circumstances?

3.6   POLICY AND LEGAL ISSUES

Getting the public involved in discussions is one thing, assuming that anyone takes notice of their views is quite another (Barker and Peters 1993). One of the most important areas to consider in public dialogue with science is what for and why are we having such dialogue and how to really affect democratic change and policy in the way technology is developed and implemented. As discussed earlier by POST, it is a waste of everyone's time unless someone is prepared to listen and do things differently as a result. The worst case scenario as we have already intimated is that public dialogue is no more than a public relations exercise to get people to agree with some contentious technology which will make the government and share holders much money. Graham (1998) argues that projects are increasingly 'participatory' with the public; however, he makes it clear that this can lead to a new type of engineer 'one who is as skilled at manipulating the public with public relations. .. Today engineers speak of 'managing the public which tends to get obstreperous with large construction projects' (page 122). This new type of engineer has learned the language of public hearings, of the courts and of community relations.'

Barker tells us that street level democracy (Barker, 1999, page 250) on its own 'is no match for the difficulties of power and livelihood at the margins of globalisation. It needs to be linked to wider organisations of interest and information and to enter into dialogue with centres of power in government and elsewhere that have an impact on the locality. …'

These authors suggest that good questions to know before taking action are as follows:
1. What's the problem?
2. Who knows what about the problem?
3. Are the experts accessible to decision makers?
4. What is the information to be used for?
5. How is the advice to be institutionalized?

3.7   GUARDIANS OF KNOWLEDGE

A major consideration in the discussions about public relations with technology is how much the lay public can understand. Who is in a position to give a judgement on whether something should be allowed to be developed, or the legalities of impacts on humans, animals and the environment? Imber discusses the problems of communicating scientific evidence to a jury or to an anxious pubic (Imber, 2002). His take seems to be one of getting the communication 'right.' The problem is where there is no clear evidence and where a balance of probabilities needs to be presented. Bias is a real issue – and we have already discussed above the problem of industrial or government interest in particular outcomes.

The same 'evidence' can be viewed in very different ways. For some, the Green Revolution was a wonderful era of chemical agricultural development supporting the poor. Not so for others.

'The Biotech ads portray a brave new world in which nature will be brought under control. Its plants will be genetically engineered commodities, tailored to customer's needs. Agriculture will no longer be dependant on chemicals and hence will no longer damage the environment. Food will be better and safer than ever before, and world hunger will disappear. .. many of us remember vividly that very similar language was used by the same agrochemical corporations when they promoted a new era of chemical farming, hailed as the 'Green Revolution'. .. It is well known today that the Green Revolution has helped neither farmers nor the land nor the consumers. The massive use of fertilizers and pesticides changed the whole fabric of agriculture and farming, as the agrochemical industry persuaded farmers that they could make money by planting larger fields with a single highly profitable crop and by controlling weeds and pests with chemicals. ..With the new chemicals, farming became mechanized and energy intensive favouring large corporate farmers with sufficient capital .. all over the world, large numbers of people have left the rural areas and joined the masses of urban unemployed as victims of the Green Revolution. The long term effects of excessive chemical farming have been disastrous for the health of the soil and for human health, for our social relations and for the entire natural environment. The simple truth is that most innovations in food biotechnology have been profit - driven rather than need - driven. For example, soybeans were engineered by Monsanto to be resistant specifically to the company's herbicide Roundup so as to increase the sales of that product. Monsanto also produced cotton seeds containing an insecticide gene in order to boost seed sales. .. Numerous side effects have been observed in genetically modified plant and animal species. Monsanto is now facing an increasing number of lawsuits with farmers who had to cope with these unexpected side-effects. ...canola seeds had to be pulled off the Canadian market because of contamination with a hazardous gene...(Furthermore) recent experimental trials have shown that GM seeds do not increase crop yields significantly. Moreover, there are strong indications that the widespread use of GM crops will not only fail to solve the problem of hunger, but on the contrary, may perpetuate and even aggravate it. If transgenic genes continue to be developed and promoted by private corporations, poor farmers will not be able to afford them, and if the biotech industry continues to protect its products by patents that prevent farmers from storing and trading seeds, the poor will become further dependant and marginalized... according to a recent report... 'GM crops are creating classic preconditions for hunger and famine' (Capra, pages 158–206).

So who decides what is the 'truth' that will be relayed to the public? Lets look at the demographic profile of science policy leaders in the US. These are the scientists who give evidence to the government on different policies (13).

So an ideal profile would be people making policy, interpreting the evidence and deciding the future of all of us, in a diverse and representative manner. However, the authority given to formal scientific knowledge and the fact that most scientists come from rather similar backgrounds conspires against this. In Table 3.1, we see a profile of US policy leaders over a twenty year period (Miller and Valdez, 2003). It is rather depressing that the number of women is small and not increasing much and that the percentage of younger leaders is reducing significantly.

We can also see from the report by Miller and Valdez that the majority of policy makers in the US believe that economic growth is related to basic scientific research. There are two assumptions here – that economic growth is a measure that we would want and that basic research will develop into innovations which will bring funding. We will return to this later.

There is at this level, almost no chance of informal knowledge being included by policy makers. In fact, the EPA (Environmental Protection Agency, 2003) Science Policy Council in the US gives the following guidelines about what constitutes appropriate knowledge. It is very clear that scientific positivism is required and no indigenous knowledge would get through the door.

> Soundness - the extent to which the scientific and technical procedures measures methods and models employed to generate the information are reasonable and consistent with the intended application
>
> Applicability and utility - the extent to which the information is relevant for the agency's intended use
>
> Clarity and completeness - the degree of clarity and completeness with which the data assumptions, methods, quality assurance, sponsoring organizations and analyses employed to generate the information are documented
>
> Uncertainty and variability - the extent to which the variability and uncertainty (quantitative and qualitative) in the information or in the procedures, measures, methods or models are evaluated and characterised
>
> Evaluation and review - the extent of independent verification and validation, peer review of the information or of the procedures, measures, methods or models

Challenge Box: Who do you think should be involved in technological policy making? Why?

3.8   THE INTERNET

The internet has not been around for a very long time, but for those who grew up after its invention and implementation, it can appear as if it has always been with us. The appearance of the world wide web had a profound impact on the way we communicate and think, and we cannot address this

Table 3.1: Demographic profile of Science policy leaders (Miller and Valdez, 2003)

	Science Policy Leaders			
	1981	1984	1986	2002
Age				
Less than 50	42%	41%	36%	12%
50 to 59 years	31	30	30	30
60 to 69 years	22	24	28	30
70 years or more	5	5	6	29
Gender				
Female	11	12	NA	18
Male	89	88	NA	82
Educational Attainment				
Baccalaureate	10	9	7	6
Masters	14	13	14	10
Law	10	7	7	1
M.D.				4
Ph.D.	66	71	72	78
Discipline				
Biological Sciences (including M.D.)	16	16	23	30
Physical Sciences	25	29	32	30
Social Sciences	19	17	15	7
Engineering and related professional	21	23	17	30
Other (including education and law)	17	15	13	3
Number of respondents	287	630	508	331

important area in any detail here. The seminal book by Turkle (1995) which although only 4 years old already seems very dated - explores in detail the idea of a character which exists only inside the net. Today, alternative personas or avatars are an unavoidable part of the young person's landscape. However, despite the intentions of the original designers, the internet is a territory only available to the privileged part of the world. Even with the attempts to create low cost computers and increase accessibility in many parts of the world there is no computer and no server in many towns, and in many developing countries, connections are so poor that huge numbers of applications are not available to them.

In their book 'The Internet and Politics: Citizens, Voters and Activists,' Oates et al. (2006) look at the definition of the role of the internet in the promotion of a civil society in advanced industrial democracies. They try to gauge whether the internet serves as a beacon of democracy or as a tool for oppression in non-free states. They describe the initial work on e-democracy and political participation which had high hopes for the betterment of society. It was considered that there would be an opening up of a decentralized interactive public space in which people or 'netizens' would form new social bonds and create new opportunities for political decision making. Others, however, were

Figure 3.13. Accessing the world.

concerned that the internet would reduce the possibility for collective action. The biggest problem first encountered as discussed above was that only a minority had access to the internet – the affluent minority. Furthermore, critics suggested that social isolation was happening rather than the desired community building – the dehumanizing effects of sitting in front of a computer all daylong. Recent work has shown that sociability of individuals has dropped and feelings of disconnection to society increased with higher use of the Internet. Oates et al. have also found that computer mediated communication and email were inferior compared with telephone even for close social relations. The team also try to analyse whether the internet serves as a particular catalyst for outsider groups (defined in this study as terrorists).

Challenge Box: How has the internet affected your life? Can you imagine a life without facebook? What about your parents?

3.9   SCIENCE EDUCATION/LITERACY

We saw earlier that policy leaders believe that the economy of the developed world is related to the science and technology base. It is clear when reading any news article about development that this is assumed to be the power house of ideas that create money.

It is also assumed that the level of science education is an indicator of potential development and progress in the future. Drori (1998), however, critiques this premise and investigates its claims. She first examines the model of science for development and declares that it rests on the theoretical foundations of 'structuralist-functionalist modernisation theory which is essentially a liberal, socio-evolutionist perspective' (page 52). This modernisation theory suggests that less developed countries (LDCs) follow the footsteps of developed countries (DCs). Science and education are thus seen as the mechanisms that proved effective in the progress of DCs. In this scheme then science education is seen as shaping positive attitudes towards modernization and training candidates in science and technology to prepare them for higher education. 'in addition to the general objective of mass education as socializing modern individuals into participatory civil and political life, science education takes the additional role of professional training and providing the local economy with skilled labour' (page 53). We have seen the effect of this way of thinking in that the amount of science and maths has increased in most schools and subjects such as art have been diminished. Drori critiques the

Figure 3.14. Flying to success with a science degree.

idea of science education for economic development. It is shown that there are newly industrialized countries (NICs) experiencing high rates of economic growth in spite of the fact that they trail other regions in the relative share of curriculum devoted to science subjects. Also, it is noted that poor countries are less likely to offer quality science education. Furthermore, there are cultural barriers that impede the transfer of knowledge (e.g., Islamic countries barring some scientific information).

Science education is furthermore accused by Drori of providing mis- information which is not applicable to local needs and channels labour to a narrow range. In Lesotho, for example, most science and engineering graduates go overseas for their careers as there are few jobs at that level inside the country. 'There is little glory or recognition for improving a 'jiko' for instance, compared to work in the area of nuclear physics. It is obvious, however, which of these efforts has more direct applicability and potential use for the people in Kenya' (Bruchhaus, 1985, page 63, Drori).

'In order for school science to have a desirable effect on technological and economic development, it is important to recognize science as a cultural form' (Medvitz, 1985). As such Drori continues 'science is not applicable to all cultural contexts. Science education policy should, thus, be cautiously implemented, since we do not hold sufficient knowledge about the ways by which cultural forms affect national development' (page 65).

Challenge Box: Drori makes some very controversial statements in this text - what are your thoughts about her views on science education?

Drori believes that it is no more than a collective faith that science education will improve progress of a nation. 'The current trends of expansion of science education are enabled only due to the fact that the *value* of science education has already been established. How is this myth globalised? 'there is a need for further examination of the historical progress by which the current discourse of science education for development was formulated. Such investigation will reveal the social groups that were and maybe still are involved in the constructions of this discourse and who serve as its careers, their power relations, their formal and informal aims, etc.' (page 71).

3.10 SCIENCE AND THE MEDIA

Science TV is developing fast. Documentaries are still the most common form of programme developed, but schemes such as the Public Awareness of Science (PAWS) scriptwriting award organised by the British Association of Science and the Royal Society in the UK in the 1990s acted as a match making agency holding drinks receptions whereby scientists and script writers would team up to create TV series which featured science. Not many successful scientific soap operas were created, but it was a start.

Building the Impossible was a four part documentary series, demonstrating many of the features of the recent trends. Produced by BBC Science (and TLC – the Learning Channel in the US) and filmed over two years, I was asked, along with Chris Wise, a Civil engineer, to form the team to take on the challenge of building four folleys which may or may not have been possible at the time. One feature of the filming that I liked very much was that the team were asked not to share thoughts and discuss solutions 'off camera' so that the public gained as much as possible from our own 'live' creativity. We learnt how to 'seal' an Egyptian tomb, we designed and built an 18th Century airship as well as a huge Roman catapult intended to destroy the walls of Jerusalem. For the very first programme, however, I had to learn how to make a submarine.

Figure 3.15. Building the impossible in Egypt.

Drebbel was born in approximately 1572 at Alkmaar in the north of Holland. He is known to have presented James I with his 'perpetuum mobile,' a form of clockwork which was an attempt to demonstrate the principle of perpetual motion. It made him famous amongst his contemporaries. One of the most intriguing of Drebbel's supposed inventions is the first ever submarine. With the developments known at the time, could the submarine have actually worked? There are no documents written about the submarine by Drebbel himself. The only information comes from second or third-hand reports; however, these were written by well-respected scientists of the day. Robert Boyle was told of the event by an eminent mathematician who had interviewed one of the rowers and Boyle himself spoke to Drebbel's son-in-law. Dutch scientist, Isaack van Beeckman, was told of the event by his father who was in London when the event took place. Constantyn Huygens, Dutch poet and ambassador in London apparently wrote of it in his autobiography. There was also a small reference in Drebbel's own notes to an experiment in which he heated saltpetre. The result of heating saltpetre is oxygen. This is a revelation because oxygen was not officially discovered for another 150 years and it was thought that this might explain how Drebbel dealt with air quality in the sub - at least to some extent. There are many other passing references and, by piecing all of them together, the general impression before we began was that it did submerge and was rowed with perhaps 12 oarsmen, from Westminster to Greenwich (an hour underwater).

Figure 3.16. Drebbel's submarine.

This was all we knew before we took on the challenge. The submarine was a huge amount of fun to make, and these programmes were extremely successful by the standards of viewer figures and feedback. Some won awards. But it is worth considering these programmes in relation to the idea of public understanding and appreciation of engineering. Do they really help the public understand the spirit of enquiry and discovery needed to develop these previously unimagined machines and mechanisms?

Recent correspondence with students at Queen indicates that public understanding through media is a clear area of interest but one which we do have quite right as yet. One student told me:

Although I realize that getting input from society is vital, I also think that society may have a lot to learn about engineering before this input is completely valuable. For example, my parents always complain that they don't know what I'm doing anymore (despite attempts on my part to explain). One culprit in this is the media. As far as I am aware, there are two popular portrayals of engineers: the omniscient Star Trek engineer and the builds-impossible-stuff TLC/Discovery Channel engineer. Both of these portrayals are disingenuous and seem to emphasize that engineering is a field replete with high-impossible tasks.

Apart from being quite amusing as he was not aware that I was the presenter on 'Building the Impossible,' I found this statement and others like it indicative of the interest that students have in being able to talk to friends and family about their subject. Aware that we have along way to go, Canada has just recently developed a Science Communication programme in the Banff Arts Centre, hosted by Jay Ingram, Discovery science show presenter, for TV and newspaper journalists as well as museum educators and theatre directors, to help them learn the art of science communication. However, one thing is clear as the student suggested – we have to move beyond the continual presentation of engineering as a series of challenges to make the fastest, tallest, most powerful …Engineering in particular is rarely presented as a way to make the world a more environmentally sound ,just, safe and peaceful place.

3.11   SCIENCE THEATRE

Science theatre is developing in a very different tradition to science TV. Shepherd-Barr (2006) suggests that science plays bring politics back to the stage as the political implications of scientific and technical developments become an important part of the drama and science playwrights have as a common interest the concept of justice. She uses Durrenmatts notion (Peppard, 1969:72 in Shepherd-Barr, 2006:50) that 'justice is a question of what to do with the power that stems from scientific knowledge.' 'Playwrights have used science to specifically probe the problem of human progress through the advancement of knowledge' she tells us (Shepherd-Barr, 2006:18). She quotes Glynne Wickham (Wickham, 1962:47,52 in Shepherd-Barr, 2006:47) as stating that ' we live in a world where the individual has committed the control of his agricultural and industrial economy, of his communications and even of his bodily and mental health to scientists' but 'because the theatre can provide a forum for the examination and discussion of the human condition its relationship with its gods and its interest in itself, collectively and individually' it might' serve as a safeguard to the tyranny of science.' Far from presenting science in a positive light we see that it is often declared as the evil destroying the world. Many successful science plays have 'ethical conflicts as well as personal ones, centering on the use of potentially lethal scientific discoveries. They pose dilemmas for the individual scientists, as well as on the much larger collective scale of all humanity, which makes their conflicts all the more powerful and dramatic' (Shepherd-Barr, 2006:41). The plays themselves therefore often focus on the individual scientist and their responsibility in the world.

As previously discussed in Michael Frayn's play Copenhagen, for example, Neils Bohr and Werner Heinsenberg together with Bohrs wife Margrethe, relive a meeting which took place in 1941 when the younger Heisenberg visited his old post doctoral supervisor and mentor Bohr in his home in occupied Denmark. No one really knows the purpose of the visit but it is the subject of much conjecture and controversy. Did Heisenberg go there to try to get Bohr to agree to a pledge of non-action as they were two of the world's most important atomic physicists, and if they decided not to create the bomb, they could influence their teams also. Or did he go there to learn how to make the bomb as Germany was falling behind in the research? Shepherd-Barr discusses Ronconi's argument that 'a play like Copenhagen does engage science deeply and meaningfully, but the science is mediated through biography. This gets the audience too wrapped up in plot and personalities.' She quotes Ronconi from an interview with Maria Grazia Gregori ' a scientific topic such as the uncertainty principle has become part of Heisenberg's biography' (Shepherd-Barr, 2006:200). In the play, Neils Bohr and Werner Heinsenberg - see Fig. 3.17.

Figure 3.17. The 2004 Critical stage/ Theatre Kingston production of Copenhagen.

Frayn tells us, despite the fact that 'I could not go back in time and hear how these people actually spoke and what their manner of being actually way ...as with fictitious characters, they at some point seem to take over and have a voice of their own' (Hampson, 2004). The tendency with these plays which have a hero or a villain is to blame the scientist for not speaking out, for not communicating or for simply being arrogant and self satisfied with their own discoveries. What works well as a piece of drama is if we see the scientist in some sort of dilemma or difficult situation as a result of his or her actions.

Shepherd-Barr suggests that 'In the journey from scientific source to newspaper, magazine or the Internet report, serious omissions, compressions and distortions can occur; not only the science gets misrepresented at times but the very process of discovery' (Shepherd-Barr, 2006:47).

She considers that the fault lies with the scientists not the journalists as they are 'not always the best spokespersons for science…there is an urgent need to make science trusted again' (Shepherd- Barr, 2006:48). She conducted an interview with Prochiantz in 2003 who told her that 'there is not enough transparency and openness about what scientists do on a daily basis 'we see nothing of the process of science or the human beings inside the scientists; we just see the end results' (Shepherd-Barr, 2006, 48). She believes science plays attempt to address this.

In the spirit of public dialogue, the role of the talk-back is seeing a revival with science plays. In this sense, the role of science plays is often considered to be about educating the audience so they can make informed decisions. In the Kingston 2004 production of Copenhagen (Critical Stage and Theatre Kingston) we hosted several talk- backs as shown in the photograph of Fig. 3.18. Peter Calamai from the Toronto Star offered to host one of these and we had a different panel of speakers at each event to spark off discussion including on one evening, a telephone appearance from Professor Lewis Elton, education professor and physics professor who had personally met Bohr and his wife Margrethe, and taken a walk with him, just as Heisenberg did in the play. Lewis produced an original letter from Bohr to Oliphant in Birmingham discussing Frisch's visit to the UK and the potential of a bomb from uranium. Another audience member produced an original letter to her mother who had worked in a minor role in Los Alamos, thanking her for her contribution to the making of the bomb. The discussions were wide ranging, some focusing on the science, some on the biographical and some on the political and historical. The most interesting outcome for me was the outrage expressed well by one audience member that 'personalities could have such a profound effect on the future of the world.'

One very common form of engagement with the public happens in Theatre-in-Education. Increasingly TIE companies are delving into scientific areas as well as the more usual moral development issues.

An example of a traditional science play in a TIE setting involved my theatre company Critical Stage (CSC) toured London schools with a play entitled 'Dr Concept Detective.' The objective of this project was to take a series of scientific concepts, develop them as a 'who dunnit' style detective dramatisation of the scientific thought process. In the play, a boy is killed because his crash helmet has split open and the task of Dr. Concept Detective (naturally a woman) was to find out 'whodunnit' – who was responsible for the faulty manufacturer of the helmet. The play focuses on the field of failure analysis, a very important area within materials science. During the play real scientific results were discussed and real materials used as props.

The workshop following consisted of the cast, most of whom were also scientists or engineers, holding small group discussions with the students about the play. It was clear that many of the important scientific issues were, in fact, picked up by the students even though they came from a very wide range of academic ability, background and interest science. It was also obvious that a whodunit style works very well in this context and that issues of scientists' responsibility and ethical behaviour can be easily discussed without moralizing. Furthermore, the precariousness of scientific evidence was brought into the forefront for students to ponder, in a culture where they are still taught

Figure 3.18. Talk-back after the Kingston production of Copenhagen.

scientific positivism in their school classrooms the inevitability of scientific discovery and the search for 'truth.'

'Experience the Innocence,' another CSC production, incorporated an example of a more experimental form of theatre where the scientific concepts themselves were enacted. It was a complex multi - component project run over two years with a twin intention of raising the profile of science within the Merseyside area and to increase the potential of women and girls becoming involved in science. In addition, it was intended to substantially enhance the basic employability of women to return to work. In the first phase of the project, a group of women were employed to take part in weekly workshops where they developed basic skills as well as learning about science and theatre. They formed a theatre company which they called 'Dream 2B Green.' They learned about playwriting and acting, as well as materials science and in particular the science of recycling plastics. They went on to create a play about recycling plastics which featured Polly Ethylene, her cat PET and her boyfriend Reece Icyle. Polly was deciding whether or not she might pass through the recycling 'gate' to go to a new world.

Figure 3.19. 'Polly Ethylene' and her boyfriend Rees Icyle, with PET the cat (Dream 2B Green).

The play, which was intended to help students learn about which plastics might be recycled and why, was taken to schools in the local Merseyside area together with a workshop involving many hundreds of plastic bottles. The women also developed skills by creating an interactive CD-ROM, two booklets: 'Student Guide' and 'Teacher's Guide,' to introduce the recycling knowledge and terms involved in the play and a website to support their production. 506 girls between 14-16 years age from Merseyside schools took part in these activities. In the second phase of the project, the women worked together with Critical Stage Company to incorporate their play into a one act play for the Edinburgh Festival 'Experience the Innocence' incorporating William Blake's poetry and perspectives on the impact of science on society which subsequently toured the UK. Talk – backs were held with the audience after each play in the Merseyside area.

Deborah Mutnick (2006) discusses the importance of dialogic forms of education and art, in which the audience are encouraged to take part in some way. She discusses in particular the work of Boal and Freire. Boal developed a very interactive form of theatre called 'Theatre of the Oppressed.' In the photograph of Fig. 3.20, classic theatre of the oppressed style 'forum theatre' is being presented at a conference on International Partnerships hosted by the University of Calgary. The audience are allowed to step into the action, facilitated by the 'joker' who invites them to do so. The Forum Theatre presented here is a representation of a community meeting between locals and

Figure 3.20. Forum Theatre at the University of Calgary.

the government. On this occasion, it was a representative of World Bank who ran in to say 'That's not how it is!'

3.12   FINAL THOUGHTS

This chapter has embraced the way in which science, technology and engineering meets people, how the government and industry worry about how people will react to certain technologies, and how to develop a better dialogue between the engineers and the users. We have explored the reason why this is a hot topic at the time of writing and that many organisations are developing ever more creative ways of improving public interaction with engineering knowledge. There is still much skepticism as to why the government of any particular country may want this. There are those who fight for the public to have better information for true citizenship and democracy. There are those that believe the public are not being better informed but better controlled in the way that they think so that they do not prevent economically beneficial projects from going forward on the basis of fear and uncertainty about potential implications. We have tried to present some more progressive ways of engaging with the public that will alleviate this latter issue and allow for a truly two way conversation, such that the public may indeed inform and influence policy but that they do so from a position of awareness and knowledge.

CHAPTER 4

Globalisation, Development, and Technology

Figure 4.1. The Global Engineer.

4.1 INTRODUCTION

In this chapter, we will explore the global context in which we find ourselves working as engineers. When we start thinking about a global context, we often come across the word 'globalisation' and 'developing countries' – also the 'third world.' Before we can understand any issues of relevance to our profession, it is important to focus on some of the key global players and to look at some of these terms in detail.

The term 'Third World' is a term used along with First World and (to a much lesser extent). Second World to divide countries into three broad socio-political and economic categories. The third world referred to those countries especially in Latin America, Africa, Oceania, and Asia, which were not aligned with either the Soviet nor American blocs during the Cold War. The term was created in 1952 by Alfred Sauvy, a French demographer. Today the more common term is developing or underdeveloped countries and the term is synonymous with all countries in the developing world, independent of their political status.

Some people disapprove of the term "developing countries" as it implies that industrialization is the only way forward, and they believe it is not necessarily the most beneficial model. They prefer the term 'Global South.'

The major players in global economics, often criticized for being too powerful and US centric, are the World Bank, the World Trade Organisation (WTO) and the International Monetary Fund (IMF).

The World Bank represents five international organisations who provide advice to countries for economic development and poverty reduction:

- the International Bank for Reconstruction and Development (IBRD), established in 1945,

- the International Finance Corporation (IFC), established in 1956,

- the International Development Association (IDA), established in 1960,

- the Multilateral Investment Guarantee Agency (MIGA), established in 1988 and

- the International Centre for Settlement of Investment Disputes (ICSID), established in 1966.

Governments can choose which of the above to sign up to. The IBRD has 184 members, whilst the others have between 140 and 176 member governments.

The World Bank is part of the United Nations system. Its governance structure is different, however, as each institution in the World Bank Group is owned by its member governments. Each one subscribes to its basic share capital, with votes proportional to shareholding - Membership giving voting rights that are the same for all countries. However, there are additional votes which depend on financial contributions to the organisation and as a result, the World Bank is controlled primarily by developed countries, while clients have almost exclusively been developing countries.

The International Monetary Fund (IMF) describes itself as 'an organisation of 184 countries, working to foster global monetary cooperation, secure financial stability, facilitate international trade, promote high employment and sustainable economic growth, and reduce poverty.' With the exception of North Korea, Cuba, Liechtenstein, Andorra, Monaco, Tuvalu and Nauru, all UN member states either participate directly in the IMF or are represented by other member states. The IMF monitors **exchange rates** and **balance of payments**, as well as offering technical and financial assistance.

The World Trade Organisation (WTO) is an international, multilateral organisation, which sets the rules for global trading. It resolves disputes between member **states**. WTO formed out of the GATT (General Agreement on Tariffs and Trades). The WTO has 150 members (76 members at its foundation with 74 members joining over the next ten years). The 25 states of the **European Union** are represented also as the **European Communities**. All three of the above organisations were formed at the **United Nations Monetary and Financial Conference** in **Bretton Woods, New Hampshire, United States**, on **July 22, 1944**.

Challenge Box: Find an article in the newspaper which mentions the IMF or World bank. Try to ascertain the 'position' (as discussed in the last section), of the journalist and of the newspaper.

4.2   THE INDUSTRIAL REVOLUTION (1760-1830)

The history of engineering is often considered to be divided into four phases:

Pre-scientific revolution: Master builders and Renaissance engineers, e.g., Leonardo da Vinci.

Industrial revolution: (1760-1830). The change from craftsmen to engineers and increase of mass production, introduction of factories.

Second industrial revolution: (1830-1930) chemical, electrical, and other science-based engineering branches developed electricity, telecommunications, cars, airplanes.

Information revolution: (1950 – present day) microelectronics, computers, and telecommunications jointly produced information technology or IT.

There are many differing views about the first Industrial Revolution. In many engineering text books, it is presented as the most wonderful era of invention and discovery - the beginning of the modern world and the creation of engineering as we know it. To others, it was the beginning of the relationship between capitalism and engineering. To the workers, it was a time of great misery.

The factory developed and increasing numbers moved to the cities to become workers. Housing in the cities was not adequate, and there was abundant child labour. Gender roles would also become defined as men would increasingly work away from home. This does not seem very different to what we call development in many parts of the world today.

The reports of the Commission (Factories' Inquiry Commission of 1833) touching this barbarism surpass everything that is known to me in this line.. The consequences of these cruelties became evident quickly enough. The commissioners mention a crowd of cripples who appeared before them, who clearly owed their distortion to the long working-hours. This distortion usually consists of a curving of the spinal column and legs, and is described as follows by Francis Sharp, M.R.C.S., of Leeds:

"Before I came to Leeds, I had never seen the peculiar twisting of the ends of the lower part of the thigh bone. At first I considered it might be rickets, but from the numbers which presented themselves, particularly at an age beyond the time when rickets attack children (between 8 and 14), and finding that they had commenced since they began work at the factory I soon began to change my opinion. I now may have seen of such cases nearly 100, and I can most decidedly state they were the result of too much labour. So far as I know they all belong to factories, and have attributed their disease to this cause themselves."

"Of distortions of the spine, which were evidently owing to the long standing at their labour, perhaps the number of cases might not be less than 300."

Engels (Conditions of the Working class in England) 2004

4.3   ENGINEERING AND DEVELOPMENT IN 19TH CENTURY

Once the industry had been developed in the UK, it was possible for the British to expand their developments to other countries. We are not going to discuss the British Empire in any detail but just look at one important period of famine and consider the role engineers played with the disaster. In 1876, the monsoon failed to arrive in Madras, and millions starved to death. Davis, amongst other scholars believe that the drought that occurred did not have to cause a famine and in fact, only has been done through the poor management of a country. India had suffered many droughts before and the local farmers were able to survive by careful management and storage of their grain.

Figure 4.2. Karl Polanyi.

What does any of this have to do with engineering? Much, Davis would say. Modernization was just happening and thousands of miles of railroad track and canal had just been created. The trains, instead of helping to improve the situation and bring food to the starving, were in fact used to take grain away to Britain. The canals which replaced well irrigation were an ecological disaster. Short term wheat growth benefited, but without proper underground drainage, the capillary action of the irrigation brought toxic alkali salts to the surface, which caused extreme 'saline efflorescence.' Canal embankments blocked natural drainage and created ideal breeding grounds for mosquitos thereby proliferating malaria. Davis tells us that the British Army engineers were amazed at the

'drought was consciously made into famine by the decisions taken in palaces of rajas and viceroys…. but with equal justice the same criminal charges could be (and were) lodged against the British administration..
…'Early and energetic organization of relief and above all, the deferment of the collection of land tax might have held mortality to a minimum" p51 Davis (2002)

Figure 4.3. Rama Varma I.

. 'The newly constructed railroads, lauded as institutional safeguards against famine, were instead used by merchants to ship grain inventories from outlying drought stricken districts to central depots for hoarding.. likewise the telegraph ensured the price hikes were coordinated in a thousand towns at once, regardless of local supply trends.. as a result of the price hike – the poor began to starve to death even in areas which were well watered' p26, Davis (2002)

Figure 4.4. Mike Davis.

skill with which previous generations had conserved water because they themselves had only made things worse (The Times, 1877, in Davis, 2002, p335).

Can we blame the engineers? Certainly, many of them were just doing their job, assigned to them by their superiors. However, we now realize that it is critical to understand our jobs and the effects that they will have on local culture, on climate conditions, on the environment, and on people's lives. We are professionals and, therefore, should be making professional decisions. It is now largely considered irresponsible to blame the boss for decisions. They may be doing terrible things but you don't have to work for them. In the next section, we will look at the situation in the world today, and we will see that engineers are still causing a lot of harm by not asking the right questions about what they are doing and why.

4.4   ENGINEERING AND DEVELOPMENT IN 21ST CENTURY

In this section, we are going to consider contemporary views on development or what is known as 'post development theory' and the implications that these have for the engineering profession. But first, our adaptation of a small story that has become notorious amongst development engineers.

One day, in a small village in a very poor country far away Army engineers had finished their work and had a few days holiday before leaving for home. During their stay in the village, they had noticed that the local women would walk a very long way each way to fetch water from the river. The engineers wanted to help the women in the village they had come to respect. They decided to dig a well with a pump in the centre of the village so that the women could get their water right outside their doors. Then they went home. Six months later they heard of a suicide in the village. The women had needed the time and space that the walk to the river provided, away from their husbands and their constant chores. Now their husbands would not stand for their chatter and time wasting by the well, in front of their own eyes. The army engineers had not thought about any negative implications of their deeds, so keen to do good.

Challenge Box: Check the labels on your clothing. Where was it made? Do you know the conditions of the workers who made your clothing?

It is not always obvious that we might cause a negative effect. Many scientists and engineers will say that they cannot be held responsible for bad effects of their work down the line. We deal with this more in Volume 2 but suffice to say that we as authors of the book think this is a poor excuse. Below we present various views of development as summarized by Leftwich (2000).

'Development seems to defy definition, although not for a want of definitions on offer. ...development is construed as 'a process of enlarging people's choices,' of enhancing 'participatory democratic processes' and the 'ability of people to have a say in the decisions that shape their lives' ..simultaneously, however, development is defined as the means to 'carry out a nation's development goals' and of promoting economic growth. Given that there is scarcely a 'Third World' dictatorship not at least in part attempt to legitimize its mandate to rule in the name of development .. it is little wonder that we are thoroughly confused by development studies texts as to what development means' (page 3, Cowen and Shenton, 1996).

Figure 4.5. Exploitation of natural resources.

Figure 4.6. Digging for gold: progress is always good.

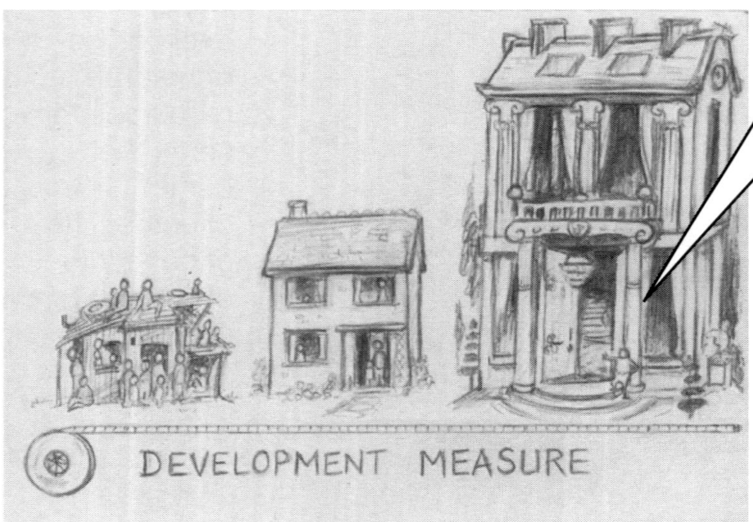

Figure 4.7. Ranking of societies.

Figure 4.8. The trustees of society.

Figure 4.9. Progressive change.

Figure 4.10. Labour power.

Figure 4.11. Thoroughly Modern Millie.

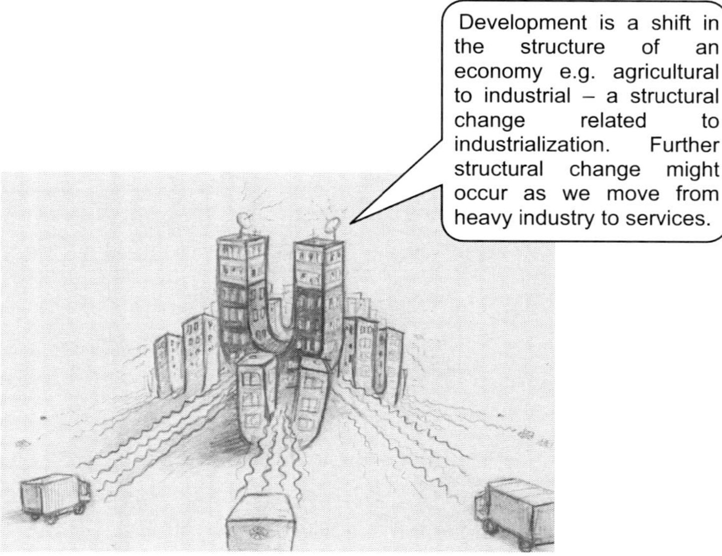

Figure 4.12. The attraction of the urban environment.

'Technological development is not automatic. It is the product of conscious and unconscious choices and decisions, all of which take place within a social and historical context. The notion that technology can be transferred without modification has few historical precedents to support it.' (Veblen in Noble, p78, 81.)

Figure 4.13. 'Actually the machines we are replacing you with can put love into the pies!'

Challenge Box: What are your views on development? Question your assumptions.

Dams are a major example of development. They offer a one-off technical solution to irrigation, flood control and power generation. Smillie (1991) discusses the development of Egypt's Aswan high Dam in 1970 (p39), which had many side effects not previously anticipated including pollution of drinking water and the destruction of the sardine industry. Furthermore, the Three Gorges Dam on Yagntze river – promised to be the worlds biggest ever power development. An 11 million study carried out through CIDA (Canadian International Development Agency) recommended go ahead. Opposition was considerable, and an organisation called Probe International obtained a copy of the report (by using the Canada Access to Information Law) and got together a panel of experts then to published 'Damming the Three Gorges; What Dam Builders Don't Want You to Know.' It dealt with issues of resettlement, flood control, design and safety and showed that economic human and ecolological costs had been under estimated and benefits overstated.

CIDA withdrew but the development is going ahead. When completed, the $25 billion Three Gorges Dam on the Yangtze River will be the largest hydroelectric dam in the world. With an installed generating capacity of 18,200 MW, the dam will span more than two kilometers across, and tower 185 meters above, the world's third longest river. Its reservoir will stretch over 600 kilometers upstream and force the displacement of more than 1.3 million people. Construction began in 1994 and is scheduled for completion by 2009. Construction on the dam itself was completed in May 2006.

Demolition in Guizhou, Central China

Smoke and dust rise after demolition efforts begin in the town of Guizhou in Central China's Hubei Province to make way for the Three Gorges Dam Project. The project has been plagued by massive corruption problems, spiraling costs, technological problems, human rights violations and resettlement difficulties. One million people have been displaced by the dam as of 2006; many are living under poor conditions with no recourse to address outstanding problems with compensation or resettlement. Said one peasant from Kai county, "We have been to the county government many times demanding officials to solve our problems, but they said this was almost impossible. They have

threatened us with arrest if we appeal for help from higher government offices." The environmental impacts of the project are extensive. The submergence of hundreds of factories, mines and waste dumps, and the presence of massive industrial centers upstream are creating serious pollution problems in the reservoir and the tributaries of the Yangtze. For five months every year when high water levels are lowered to accommodate the summer floods, a festering bog of effluent, silt, industrial pollutants and rubbish will remain in the previously submerged areas. This will create a breeding ground for flies, mosquitoes, bacteria and parasites, threatening the health of surrounding populations. Despite protests by Chinese citizens and media scrutiny of the project's impacts, private banks and export credit agencies have provided considerable financial support for the Three Gorges Dam. IRN has worked to call attention to the project's enormous environmental and social impacts and to lobby financial institutions to refrain from supporting the project.'
http://www.irn.org/programs/threeg/

Whether we see development as growth, domination, exploitation, inevitable or forced, we have little choice but to understand as much as we can about it and learn about what part we are playing. Many engineers inadvertently join in with some role of development that they didn't really understand they were playing. Engineers do many good things which contribute to a cleaner, safer world and for that we should be proud. However, engineers also for example, make the machines which are used in sweat shops. Engineers create the machines which are used to replace workers and so create more unemployment. Engineers build factories which cause health and environmental hazards. Engineers create factories in towns and cause villagers to move away from their families to find work. Engineers contribute more than possibly any other profession (apart from business itself) to the movement of capital throughout the world which causes all developing countries to become increasingly market driven. These are what we need to guard against, are responsible for and can do something about.

Schumacher introduced the notion of small is beautiful in 1955 and in 1961 came into contact with some of India's leading Gandhians. He began to articulate Intermediate Technology. The group of followers incorporated itself as a nonprofit company and called themselves ITDG (Intermediate Technology Development Group) – today called 'Practical Action.' It was intended to create appropriate technology - non capital intensive, small in scale, simple, non violent, with no negative social or environmental side effects. Some insist on a clear distinction between Appropriate Technology and Intermediate Technology. Smillie (1991) suggests that Intermediate Technology simply stands somewhere between what is known and the modern whereas the term Appropriate Technology often had strict rules applied to make it 'appropriate.' Actors in this area, he suggests became rather dogmatic to the point that none of the things Schumaker worked on would actually have passed the test. By 1980, there were 1000 institutions with an appropriate technology focus but weaknesses appeared. Smillie believes that The Appropriate Technology movement seemed to develop an uncritical coalition with those who had become disillusioned with mainstream industrial and technological development and who wanted an alternative lifestyle. 'Henry David Thoreau did not brave the harsh frontier. He may have lived a bucolic life in the Massachusetts woods, but even

Figure 4.14. Local appropriate technology being developed.

in 1845, the woods were not far from civilization and the pond was not far from his mothers house when he often had dinner when his own larder was bare..' Smillie believes Thoreau, for instance, was more concerned with trees than people. This is ever more true today where we often see alternative lifestyle, 'off the grid' environmentalists believing that they have the same aims as those who wish to make changes to existing structures to alleviate social and environmental problems.

Figure 4.15. Housing in Lesotho.

John Parry, one of the original board of directors of ITDG formed a construction materials groups in 1982 which is still active (Smillie, page 156).

http://www.parryassociates.com/

Due to a huge demand for inexpensive, durable, roofing materials, Parry created a natural fibres concrete alternative. He decided to go into equipment production when he found that the required equipment was not being made properly in Ghana and the tiles would not fit together, etc. Some criticised him, believing that the manufacture of equipment should be in the third world, but his motives were not for profit. He trains people to use the equipment in their home country. One air hostess from Ghana trained in between flights before investing in the equipment.

Challenge Box: Can you find any other examples of engineers who are actively trying to use their skills to support justice, i.e., changes which will permanently bring more equity to the people they are working with, rather than charity which helps them only short term.

4.5   GLOBAL VIEWS OF GLOBALISATION

Figure 4.16. The global marketplace.

There are many different views about what globalisation means as there are countries involved and we present some summaries of Brawley (2003). More detail can be found in his text.

Furthermore, there are many different lenses through which we might look at globalisation even if we agree on the same definition.

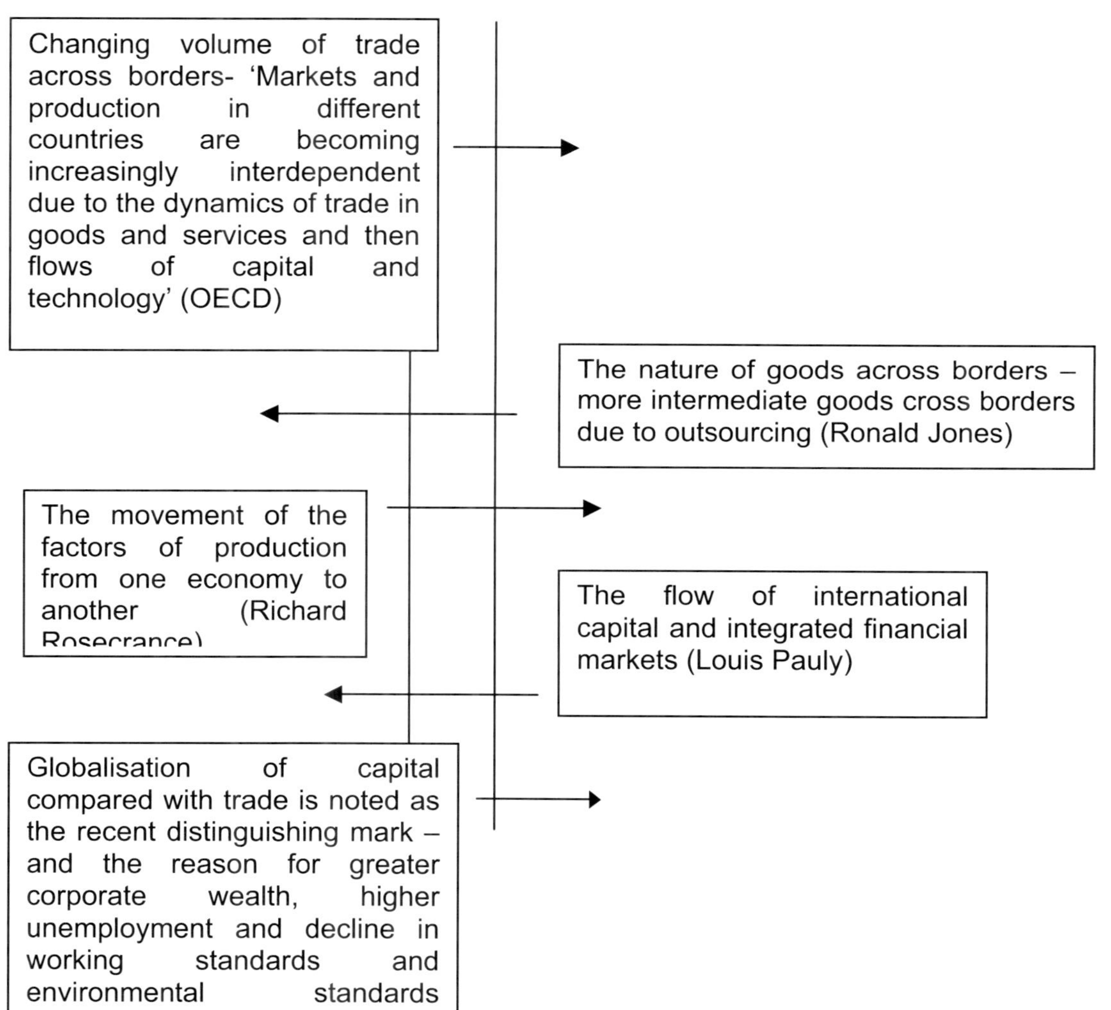

Figure 4.17. (a) Different views of globalisation.

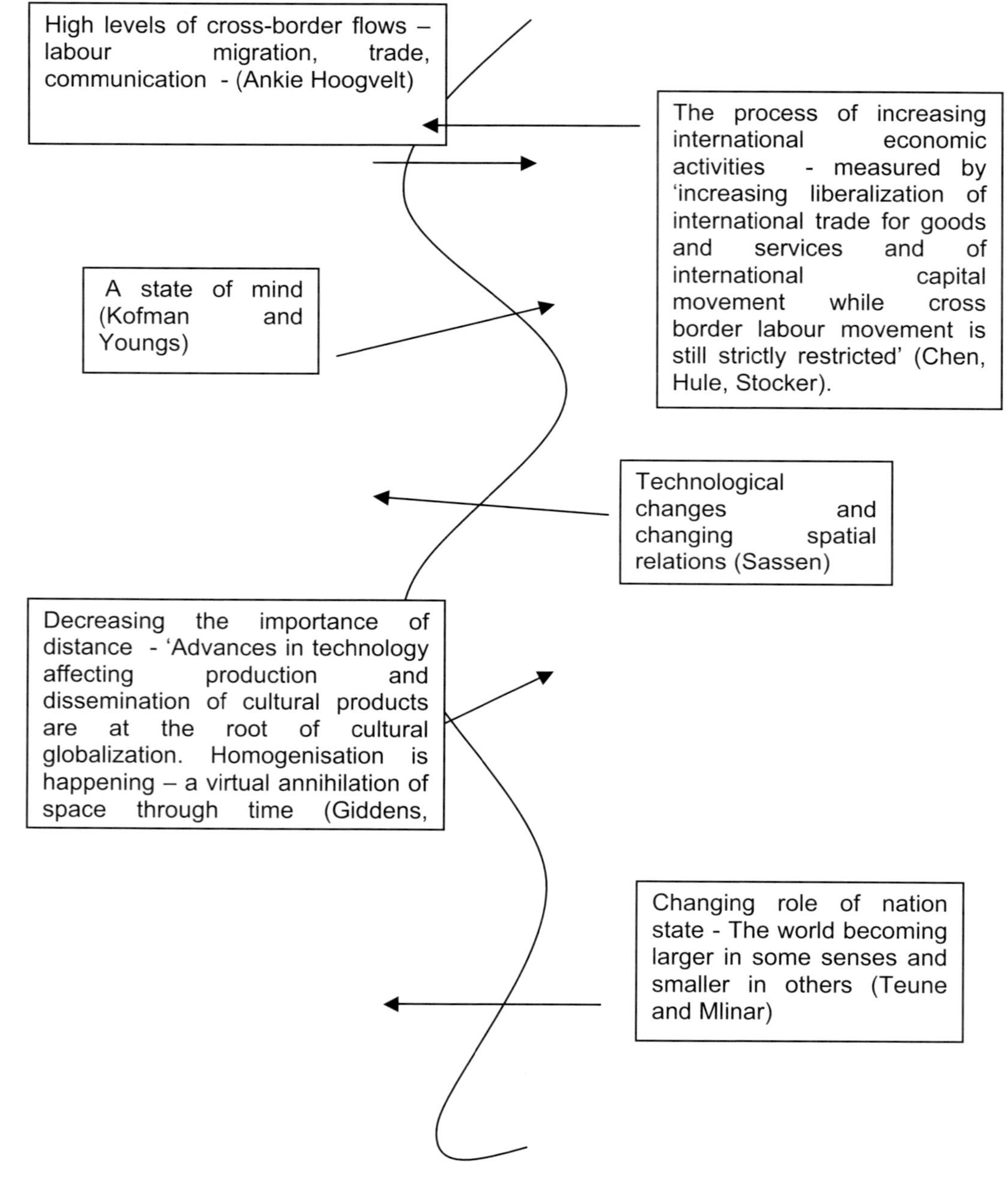

Figure 4.18. (b) Different views of globalisation.

Globalisation as progress of markets

Opening up a country to international competition gives it the opportunity of political and economic benefits. But there are concerns about state versus private sector control. How much control should there be for corporations? How much do they realize their responsibility in the care of a society?

Globalisation as creative destruction

Capital must not only have open exchange with noncapitalist societies or only appropriate their wealth; it must also actually transform them into capitalist societies themselves.

Globalisation as co-operation and interdepedence; institutionalism

States can interact and benefit mutually through globalisation

Globalisation and Democracy

An increasing number of decisions are made through intergovernmental organizations such as the World Trade Organisation or the European Union and are not democratically decided.

Globalisation and a green perspective

The green view reflects concerns about global acts which affect local issues. They concern themselves not only with environmental issues but workers rights

Figure 4.19. Lenses with which to view globalisation.

Naomi Klein is well known in Canada as a journalist and activist who works with an anti-globalisation movement. Actors in this movement are not suggesting that there shouldn't be any trade taking place across borders, nor that we should be nationalistic; both of these are common misconceptions. Imagine you live in England. Two different people might recommend you avoid buying a jacket made in China but that you buy one 'made in England.' One person might mean that they don't agree with the conditions under which workers in that particular clothing company work in China. The other may simply want to build England's economy for selfish nationalistic reasons. Others might want to support local small businesses. Globalisation to those who fight it is explicitly neo-liberal with the aim of developing Western market superiority above and beyond all else.

> **Challenge Box:** Watch the video 'Life and Debt' by Stephanie Black and consider the position of the film makers. Compare this with your own and that of your parents.

4.6   GLOBAL ECONOMIC ISSUES

Reader (2007) points out that you could be forgiven for assuming that the market system which exists today has always been present. Some say that competitiveness is in our human nature. However, if we start to look back only a few years we can see some major changes to what is called 'The New Economy.' This has emerged with three distinct features (Fig. 4.20).

Companies operating in this new environment have a number of strategies to survive (Fig. 4.21).

Work in the New Economy

Sennett (page 12, Reader, 2007) believes that the shift from managerial to shareholder power has changed the dynamics of companies. The driving force becomes one of profit – there is no real interest or concern for the individual company. Senett calls this 'impatient capital' - where companies are under pressure to present themselves in the best possible (profitable) light – appearing to be dynamic and flexible - also centralising management structures and focusing control at higher levels. This has led to more IT and replacement of some workers with computers. We also see a loss of craftsmanship and security.

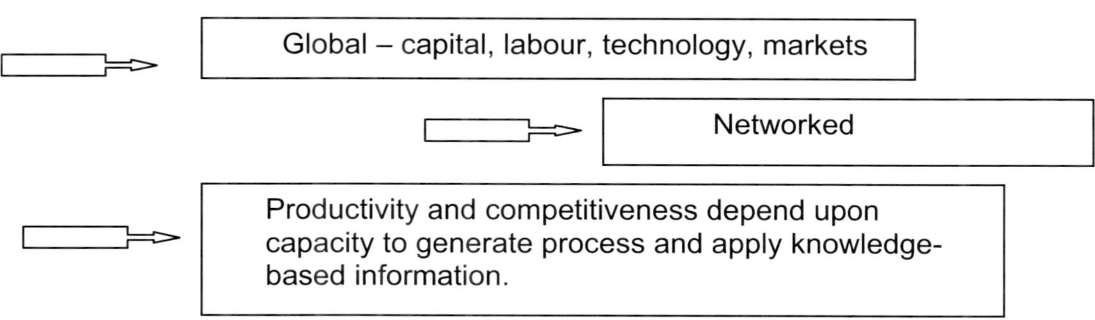

Figure 4.20. Three features of the new economy (Reader, 2007).

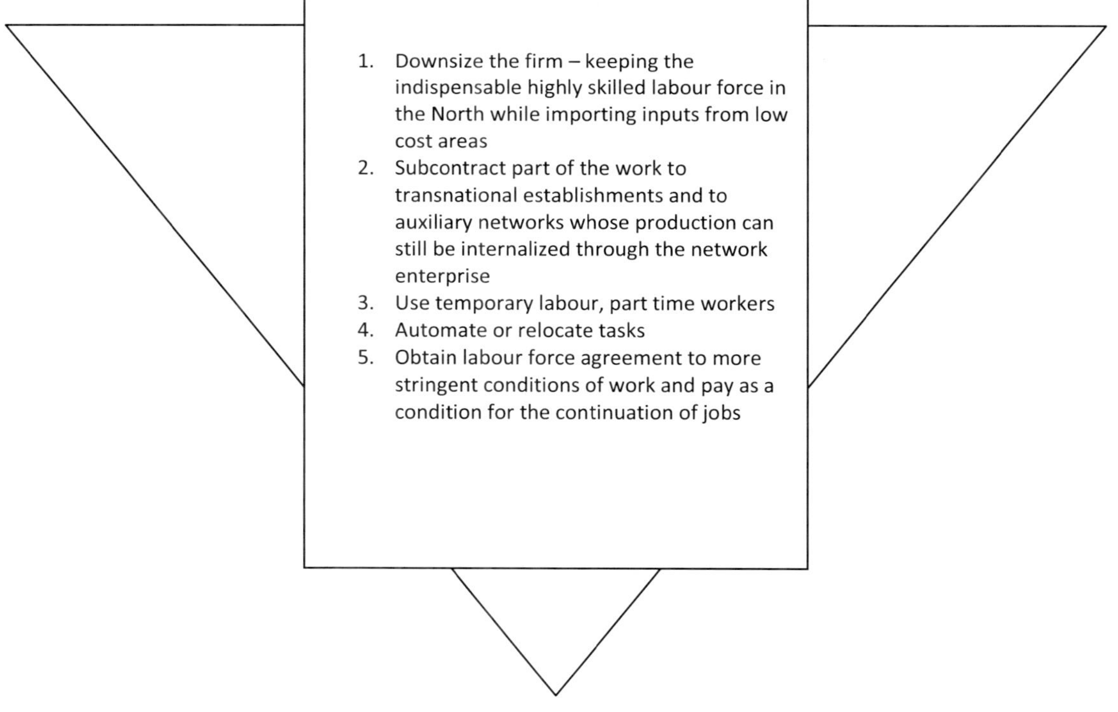

Figure 4.21. Strategies for survival in the new economy (Reader, 2007).

Figure 4.22. Impatient capital.

Competition

In 2004 (Reader, page 17), during the course of only 3 months, 58 US companies, 55 European companies and 33 from other Asian companies planned to move to China. 400,000 jobs moved to other countries. Globalisation is sometimes used as an excuse to concentrate power within fewer businesses on the grounds that this is the only way to remain competitive (e.g., US with China) as only larger companies can survive etc. Large corporations become predatory giants – mergers and buyouts abound.

Figure 4.23. Competing to survive.

The Economics of Development

The Pearson Report, 1969, was the first major report to consider under-developed countries although it is maligned by many as neo- colonialist as it ignored women, viewed poverty largely in terms of employment and simplified the idea of technology transfer. The Pearson report did recognize the problem of urbanization and malnutrition but saw solutions largely in terms of increased investment in jobs. Since then many seminal studies have been made (Smillie, 1991). In 1990, the World Bank World Development report devoted itself to poverty. Despite the intentions of these organisations initially, it was clear that the number of poor had increased in both real and relative terms. IMF and World Bank structural adjustment loans were intended to reduce the impact of the external

Figure 4.24. Smillie.

shocks by promoting more durable economic structures and sustained economic growth. Currency devaluation, reductions in government spending were enforced upon countries and trade and exchange liberalization, industrialization and diversification into the export of manufactured goods encouraged. However, it was assumed that there would be good access to external markets and for this to happen trade liberalization and reduction of tariffs and quotas by industrialised countries were critical elements. Smillie tells us that it was this element and not the conditions imposed on developing countries – that met with failure.

Smillie uses cotton as his good example. A natural evolution of cotton would be diversification into textiles as happened in the Industrial Revolution. Northern countries, however, invented a mechanism that worked against this Southern growth industry. Known as the Multi-Fibre Arrangement (MFA), it works directly against the WTO principle of non-discriminatory trade liberalization by restricting Southern textile imports into the North. 'It is estimated that the gain to developing countries of the removal of quotas and tariffs under the MFA would have been 11.3 billion, roughly one third of all official development assistance from all OECD countries in 1987. In Bangladesh, 700 export clothing factories had developed by 1985, and in the year that followed, US, Canada, Britain and France imposed quotas which resulted in closure of 500 factories'(page 14).

Smillie goes on to explain that conventional economic theory of development since the Second World War has been dominated by the concept of growth. 'It posits that there is a stage in a country's history during which the required conditions for sustained and fairly rapid growth must be consolidated, following which such growth is more or less assured' (Smillie, page 34). Growth rates depend upon the amount of capital investment in infrastructure and industry. The idea is that more investment in more productive sectors will create higher growth and speed in the development process. However, we also know that 'growth must also be accompanied by poverty alleviation schemes' …yet 'most emphasis in development spending is on growth…Northern commercial bank lending disappeared in the 1980s and structural adjustment lending and IMF stabilization programmes became conditional on opening economies to more rather than fewer competitive forces' (Smillie, page 34).

Challenge Box: Try talking to your friends about economic growth. Note the assumptions that are made. Many of us think within the 'hegemony' (ways of thinking of the dominant system or class) that a market driven capitalist economic system is the only feasible way to progress. In fact, we have only been thinking like this for a relatively short period of time.

'In its initial conception, the IMF was based on a recognition that markets often did not work well – that they could result in massive unemployment and might fail to make needed funds available to countries to help them restore their economies. The IMF was founded on the belief that there was a need for *collective action* at the global level for political stability. The IMF is a public institution established with money provided by taxpayers around the world. This is important to remember because it does not report directly to either the citizens who finance it or those whose lives it affects. Rather it reports to the ministries of finance and the central banks of the governments of the world.' However, ' the IMF has changed markedly. Founded on the belief that markets often worked badly, it now champions market supremacy with ideological fervour… In many cases, the Washington Consensus policies, even if they had been appropriate in Latin America, were ill-suited for countries in the early stages of development or transition. Most of the advanced industrial countries – including the United States and Japan – had built up their economies by widely and selectively protecting some of their industries until they were strong enough to compete with foreign companies. ..Forcing a developing country to open itself up to imported products that would compete with those produced by certain of its industries.. can have disastrous consequences – socially and economically. Jobs have systematically been destroyed – poor farmers in developing countries simply couldn't compete with the highly subsidized goods from Europe and America – before the countries' industrial and agricultural sectors were able to grow strong and create new jobs. Even worse, the IMF's insistence on developing countries maintaining tight monetary policies has led to interest rates that would make job creation impossible even in the best of circumstances. And because trade liberalization occurred before safety nets were put into place, those who lost their jobs were forced into poverty.' (Stiglitz, p17)

Figure 4.25. Stiglitz.

Engineering and Globalisation

According to many, engineering is one of the main driving forces of development and thereby global-isation. 'Because technology is such an important part of the growth process, aid can be of invaluable assistance in the transfer of technology from the North to the South.' However, Brawley expresses the following affects of technology which demonstrate the profound ways in which technology can impose changes to the culture of a civilisation:

Altering Public Consciousness

'advances in technology affecting the production and dissemination of cultural products are at the root of most arguments about cultural globalization.' (Brawley, p28). He suggests that 'the culture being spread around the globe is generated by corporate products and corporate decisions' and he quotes Benjamin Barber 'a culture of advertising, software, Hollywood movies, MTV, theme parks and shopping malls hooped together by the virtual nexus of the information superhighway closes down free spaces, such a culture is unquestionably in the process of forging a global *something*: but whatever it is, that something is not democratic' (p29).

Figure 4.26. The new world of theme parks and Hollywood.

Altering the Costs of Choices

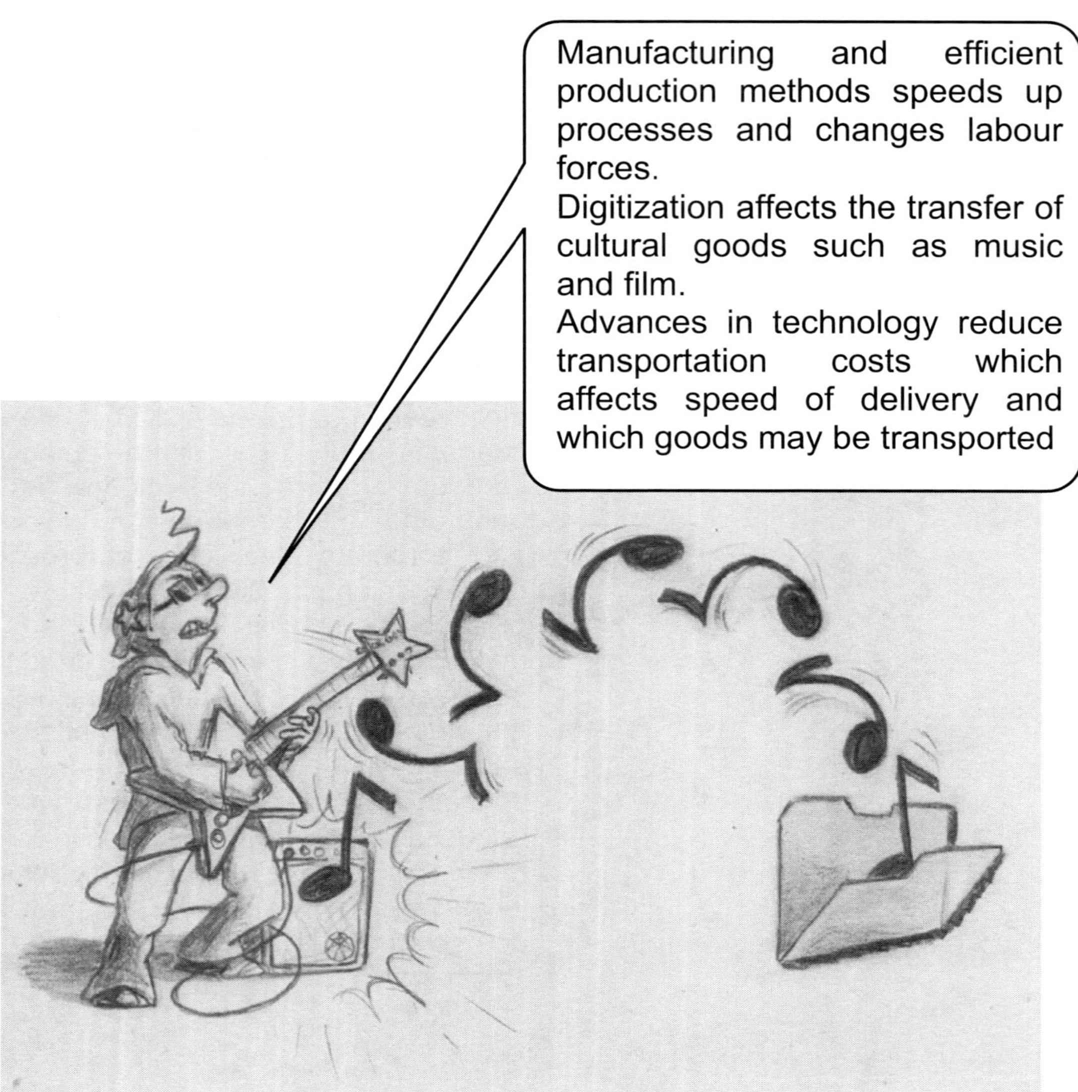

Figure 4.27. A digital world.

4.7   FINAL THOUGHTS

There are many ways in which engineering affects globalisation, and we have already considered some of these. As with our discussion of development above, it is clearly possible to say that engineers are no worse than anyone else, so why pick on them? This would be to miss the point. If you believe that we all have a responsibility to make life less intolerable for the most vulnerable populations in any country, then all professions should have education books which help their students learn about the implications and contributions that their profession is making.

> **Challenge Box:** What questions remain after reading this chapter? Write them down and keep them somewhere until they crop up again. Note how your thinking changes over time.

Part II

Engineering: Decisions in the 21st Century

CHAPTER 5

Making Decisions in the 21st Century

5.1 INTRODUCTION

Figure 5.1. Leopold.

> *We all strive for safety, prosperity, comfort, long-life, and dull-ness. The deer strives with his supple legs, the cowman with trap and poison, the statesmen with his pen, the most of us with machines, votes and dollars, but it all comes to the same thing: peace in our time.*
>
> *Aldo Leopold, A Sand County Almanac[a].*

[a] Aldo Leopold (1887-1948) is considered the father of wildlife ecology. He was a renowned scientist and scholar, exceptional teacher, philosopher, and gifted writer. It is for his book, A Sand County Almanac, that Leopold is best known by millions of people around the globe. Published in 1949, shortly after Leopold's death, A Sand County Almanac is a combination of natural history, scene painting with words, and philosophy. It is perhaps best known for the following quote, which defines his land ethic: "A thing is right when it tends to preserve the integrity, stability, and beauty of the biotic community. It is wrong when it tends otherwise."

Engineers make choices in nearly all aspects of their work. This text aims to provide you with the ability to make choices in an informed, reflective way. Here we are not concerned with choices for lunch menus or for after dinner engagements but rather making decisions concerning some of the most challenging dilemmas you will likely encounter throughout your careers. As we move farther into the 21st century, engineers will become more directly involved in issues of conflict, development and environmental sustainability. We shall confront those issues head on and offer a variety of frames of reference whereby you as an engineer can make choices whether the issue deals with some small aspect of a design or to work on a project at all. The frames of reference for decision-making include traditional approaches used in engineering throughout the modern era as well as new and exciting ideas which have just recently been applied to the professions.

Engineering is a profession with an important ethical dimension (Davis, 1998). First, we shall explore the significance of ethics for our profession, paying careful attention to the importance of codes of ethics adopted by various engineering societies. Next, as a mechanism for introduction of the complexity of issues that you may confront as engineers in the 21st century, we shall introduce

> *Can we talk of integration until there is integration of hearts and minds? Unless you have this, you only have a physical presence, and the walls between us are as high as the mountain range.*
>
> *Chief Dan George[a].*

[a]Chief Dan George (1899-1981) is best known for his writing and acting careers. His most famous roles were as the aging Chief, opposite Dustin Hoffman, in "Little Big Man," and the aging Chief opposite Clint Eastwood in "Josey Wales." He used his writing and media roles to try to give a more accurate depiction of American Indian beliefs and values.

Figure 5.2. Chief Dan George.

and discuss three cases: conflict and peace, poverty and development and environmental degradation and sustainability. These cases set the stage for our in-depth look at engineering ethics as it has been applied in the past and presently and offer you several new ideas for handling even more complex situations in the future. The new approaches include an ethic based on freedom, one on chaos, one on a morally deep world view, one consistent with a global ethic and lastly, an engineering ethic based on love.

Our goal is to enable you to consider the various issues from a wide-range of perspectives and thus ultimately be able to make your choice consistent with your deepest held values and convictions. In the spirit of the Diggers[1] from the 1960's, we are offering new 'frames of reference' from which you can consider your decisions. The Diggers focused on promoting a new vision of society free from many of the trappings of private property, materialism and consumerism. We hope to offer you a new vision of engineering which takes into account many of the elements of our society and our planet which have been historically ignored. Let us get started!

5.2   INTRODUCTORY NOTE

As you progress through the text, you will be routinely confronted with the following two items: first, we hope you come to recognize and identify with this engineering student as she progresses through the course material; and secondly, accompanying this sketch will be a 'Challenge Box' which

[1] The Diggers were one of the legendary groups in San Francisco's Haight-Ashbury, one of the world-wide epicenters of the Sixties Counterculture which fundamentally changed American and world culture. The Diggers took their name from the original English Diggers (1649-50) who had promulgated a vision of society free from private property, and all forms of buying and selling. The San Francisco Diggers evolved out of two Radical traditions that thrived in the SF Bay Area in the mid-1960s: the bohemian/underground art/theater scene, and the New Left/civil rights/peace movement.

will provide you with questions that will challenge your understanding of the material presented and also call on you to make connections to your own life and experiences in the world.

| | Challenge Box: Describe the universe. Give two examples! |

5.3   USING ETHICS CASES

The use of case studies has a distinguished history in law and business schools, and it has been very successful in the more recent emergence of medical ethics as an area of study. One of the major reasons the use of case studies is so successful is that ethical inquiry begins with problems that professionals can expect to have to face. This is in contrast to beginning at a highly theoretical level and only later considering how rather general principles and rules might apply to actual situations. By using realistic cases, students can immediately appreciate the relevance and importance of giving serious thought to ethics. Careful reflection on the cases will itself suggest the need for moving to a more theoretical level.

Purposefully and deliberately, there is a significant difference in the types of cases that we have chosen for inclusion in the present text. Many outstanding texts and other resources in engineering ethics already exist[2] [3]. Most tend to focus on either micro- and macro-ethics or a second approach described as the "tragic case" methodology (Liaschenko et al., 2002). Micro-ethics focuses upon issues related to professionalism such as integrity, honesty and reliability, risk and safety and responsibilities as an employee. Macro-ethics broadens the coverage to include issues related to the impact of engineering on the environment or the societal context of engineering. An example of a typical case in micro-ethics would be one provided by Harris et al. (2008) in which a manufacturing

---

[2] Online Ethics Center at National Academy of Engineering: As part of the new Center for Engineering, Ethics, and Society at the National Academy of Engineering, the mission of the Online Ethics Center at the National Academy of Engineering is to provide engineers and engineering students with resources for understanding and addressing ethically significant problems that arise in their work, and to serve those who are promoting learning and advancing the understanding of responsible research and practice in engineering. (www.onlineethics.org)

[3] Science and Engineering Ethics is a multi-disciplinary journal that explores ethical issues of direct concern to scientists and engineers. Coverage encompasses professional education, standards and ethics in research and practice, extending to the effects of innovation on society at large. (www.springer.com/philosophy/ethics/journal/11948)

engineer is invited by a vendor to play golf at an exclusive, private club. An example of a macro-ethics case has been offered by Herkert (2008) in his critique of the ethical responsibilities of engineering with respect to global climate change.

According to Allenby (2005), these two major approaches suggests a third. First, Allenby makes a differentiation between "micro-ethics," or ethics at the individual level, and "macro-ethics," or ethics at a group or professional level. Second, he asserts "a principal result of the Industrial Revolution and concomitant demographic, economic, technological and social changes is a planet where the dynamics of most major natural systems are increasingly shaped by human activity." The journal *Nature* suggests that we have now entered the *Anthropogenic Age* or the *Age of Man* (2003). In conclusion, Allenby argues that there is a gap between our current ethical systems and the world we have created, that it is a major gap, and that addressing it requires the serious and integrated efforts of the technical, scientific, and philosophic communities. It is that very gap which has served for the catalyst for the approach we have taken in this text.

The Worldwatch Institute has identified the three most important issues for the 21st century, which are the issue of conflict vs. peace and security, the issue of poverty vs. development and the issue of environmental deterioration vs. sustainability (Worldwatch, 2007). Accordingly to begin our discussion of making engineering decisions we introduce a set of cases which link to the Worldwatch Institutes identified issues. The case involving conflict vs. peace explores the ethical issues related to landmines. Considering poverty vs. development, we shall highlight the aftermath of Hurricane Katrina on the residents of New Orleans while our example for reflecting upon the ethical responsibilities associated with the environment centers on the rapidly disappearing bumblebee colonies. These three cases then set the stage for what is to follow.

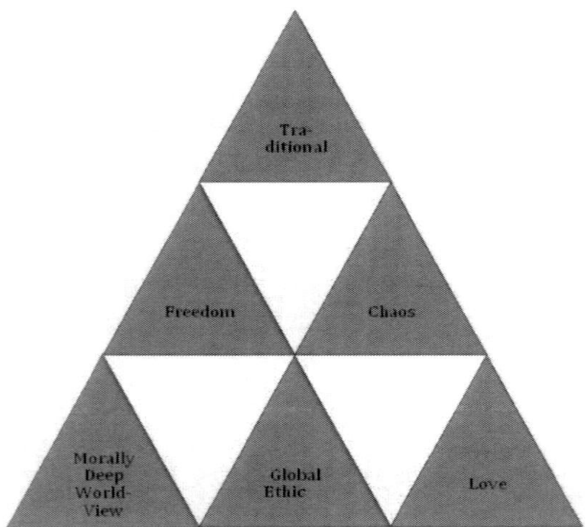

Figure 5.3. Approaches to solving ethical problems.

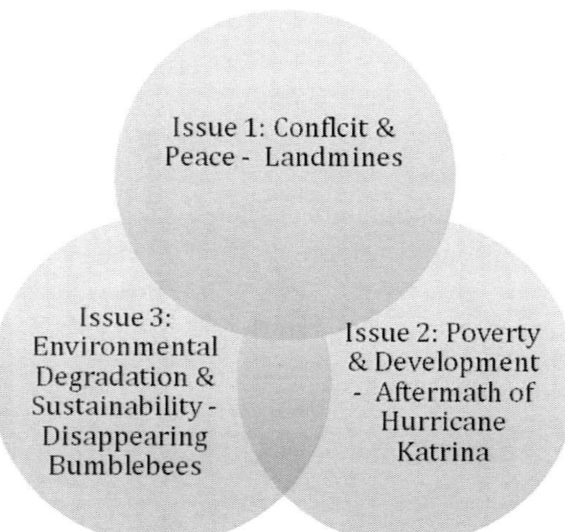

Figure 5.4. Specific ethical cases to be explored using various applied ethical approaches.

5.4   ENGINEERING PROFESSION

According to the Accreditation Board for Engineering and Technology[4], engineering is the profession in which a knowledge of the mathematical or physical sciences gained by study, experience and practice is applied with judgment to develop ways to utilize, economically, the materials and forces of nature for the benefit of mankind. Like the fields of medicine and law, engineering is considered a profession. But what is a profession, and how is it characterized? In a very general sense, a person's occupation is often referred to as his or her profession. In a stricter sense, however, a profession is characterized by special qualifying characteristics. According to Webster's Unabridged Dictionary (2005), a profession requires specialized knowledge and often requires considerable preparation in skills, methods, and principles[5]. Professions also maintain high standards of achievement and conduct, encourage life long learning, and engage in public service.

Engineers and scientists take many mathematics, chemistry, and physical courses in common in their undergraduate programs. However, their paths diverge as they begin to take courses in their majors, with engineering majors taking courses that are more applied and science majors taking

---

[4] ABET, Inc., is the recognized U.S. accrediting agency of college and university programs in applied science, computing, engineering, and technology. ABET was established in 1932 and now consists of 28 different professional and technical societies representing the fields of applied science, computing, engineering, and technology.

[5] A profession is an occupation, vocation or career where specialized knowledge of a subject, field, or science is applied. It is usually applied to occupations that involve prolonged academic training and a formal qualification. Professional activity involves systematic knowledge and proficiency. Professions are usually regulated by professional bodies that often develop and offer examinations to assess competence, act as a licensing authority for practitioners, and enforce adherence to an ethical code of practice. (http://en.wikipedia.org/wiki/Profession)

more courses in the basic sciences, being more interested in discovering knowledge with less regard to its immediate application.

On the job, engineers create tangible devices or systems such as computers, engines, and other products that perform a function. Engineers may also be involved with the process of designing systems such as satellite communications systems and robotics control systems, or with the development of a practical solution to a meet human need. Scientists also solve problems and may develop new products, but the emphasis is on the science rather than the product.

Challenge Box: Think of five professions in addition to engineering. Compare and contrast the professions in terms of their actual and potential impacts both positive and negative on society and on the environment.

CHAPTER 6

Ethics

Figure 6.1. Ethical dilemmas.

> *If it is not right do not do it; if it is not true do not say it.*
>
> *Meditate often upon the bond of all the Universe and their mutual relationship. For all things are in a way woven together and all are because of this dear to one another; for these follow in order one upon another because of the stress movement and common spirit and unification of the matter.*
>
> *Marcus Aurelius*, **Meditations** (Aurelius, 2002).

Figure 6.2. Aurelius.

6.1   INTRODUCTION

Mrs. Applebee pulled down the shade
And retired to her cozy bed
Officer Warner, round the corner
Sippin' coffee and a makin' time
He never heard the sounds of violence
Emanating from the crime
April Stark was in the park
Conducting business from her usual spot
She was witness to what happened
But conveniently forgot
Funky Wheeler, he's no squealer
He flipped his quarter high into the air
He saw their faces, but told detectives,
"I don't know nothing Jack nor do I care."
Mary's mother asked the question
"Didn't anybody hear?
When my daughter cried out "Help me!"
Didn't anybody care!?"
All's quiet on West 23rd
Nobody saw, nobody heard
All's quiet on West 23rd
Nobody saw, nobody heard
Lyrics by Julie Budd

[Reprinted with permission from the March 27, 1964 *New York Times*. Copyright© 1964 by the New York Times Co.]

37 Who Saw Murder Didn't Call the Police Apathy at Stabbing of Queens Woman Shocks Inspector

By *Martin Gansberg*

For more than half an hour thirty-eight respectable, law-abiding citizens in Queens watched a killer stalk and stab a woman in three separate attacks in Kew Gardens. Twice the sound of their voices and the sudden glow of their bedroom lights interrupted him and frightened him off. Each time he returned, sought her out and stabbed her again. Not one person telephoned the police during the assault; one witness called after the woman was dead.

In March, 1964, a New York City woman named Catherine "Kitty" Genovese was raped and stabbed to death as she returned home from work late at night. According to a newspaper report published shortly thereafter, her neighbors looked on from their bedroom windows for over 30 minutes as the assailant beat her, stabbed her, left her and returned to repeat the process two more times until she died. No one lifted a phone to call the police; no one shouted at the criminal; no one went to the aid of the young woman. Finally, a 70 year old woman called the police. Police officers arrived two minutes later but by that time the young woman was dead. For the next hour while authorities awaited the arrival of a ambulance, only one other woman came out to testify. At that time, many residents came out of their homes. The New York Times reporter at the scene of the crime reported that 38 people had witnessed some or all of the attack, which took place in two or three distinct episodes over a period of about a half hour. Notwithstanding the large number of witnesses, no one did anything to stop the attack; no one even reported it to the police until the woman was already dead. When asked why no one did anything, the responses ranged from "I don't know" to "I was tired" to "I was scared." Although the murder itself was tragic, the nation was even more outraged that so many people who could have helped seemingly displayed callous indifference.

The Genovese incident is one of the most highly discussed examples of onlookers remaining indifferent to the plight of a victim of a violent crime. Recently, another incident in Cleveland, Ohio, is a reminder that such occurrences are still commonplace. On June 28, 2008, a group of teenagers beat a homeless man to death as passers-by slowed to watch the attack, some of which was caught on video tape. The homeless man, Anthony Waters, 42, suffered a lacerated spleen and broken ribs and died at a nearby Cleveland hospital later that evening. The tape showed multiple cars slowing to watch the teens attack Waters. No one stopped or offered any assistance or called the police. The attackers appeared to be between the ages of 14 and 17, robbed Mr. Waters of a music player and headphones.

Such incidents are not confined to the U.S.; Italian newspapers, an archbishop and civil liberties campaigners expressed shock and revulsion after photographs were published of sunbathers apparently enjoying a day at the beach just meters from where the bodies of two drowned Roma[1] girls were laid out on the sand. Italian news agency ANSA reported that the incident had occurred on Saturday at the beach of Torregaveta, west of Naples, southern Italy, where the two girls had earlier been swimming in the sea with two other Roma girls. Reports said they had gone to the beach to beg and sell trinkets.

Local news reports said the four girls found themselves in trouble amid fierce waves and strong currents. Emergency services responded 10 minutes after a distress call was made from the beach and two lifeguards attended the girls upon hearing their screams. Two of them were pulled to safety but rescuers failed to reach the other two in time to save them. The website of the Archbishop of

---

[1] There are more than twelve million Roma located in many countries around the world. There is no way to obtain an exact number since they are not recorded on most official census counts. Many Roma themselves do not admit to their true ethnic origins for economic and social reasons. The Roma are a distinct ethnic minority, distinguished at least by Rom blood and the Romani, or Romanes, language, whose origins began on the Indian subcontinent over one thousand years ago. No one knows for certain why the original Roma began their great wandering from India to Europe and beyond, but they have dispersed worldwide, despite persecution and oppression through the centuries.

Naples said the girls were cousins named Violetta and Cristina, aged 12 and 13. Their bodies were eventually laid out on the sand under beach towels to await collection by police. Photographs show sunbathers in bikinis and swimming trunks sitting close to where the girls' feet can be seen poking out from under the towels concealing their bodies. A photographer who took photos at the scene told CNN the mood among sunbathers had been one of indifference.

> **Challenge Box:** Suppose you were witness to a crime as horrific as the Genovese or the Waters case. What would you have done if: (1) the victim was a complete stranger; (2) the victim was a neighbor you recognized; (3) a friend; (4) a member of your immediate family; or (5) a neighborhood dog or cat? Are your answers different? Why or why not? How did you decide what you would do? Did you use any principles? Suppose there were two assailants and each was heavily armed? Would that change your responses? Why or why not?
>
> Or suppose you were sun bathing on an exclusive beach in the French Riviera or some other exotic place. That very day you were accosted by a roving band of gypsies and they managed to have stolen your wallet/purse with all your traveler's checks. You were able to quickly get a refund and continue on with your vacation. While visiting the nearby beach, sipping a delicious alcoholic concoction and working on your tan, you notice that the several members of the same band of gypsies had tried to swim and due to strong currents had drown and their bodies were laid out on the beach close to you, covered by a blanket. How would you feel? What would you do if anything?

The Genovese, Waters and Roma children tragedies suggest a range of questions for each of us. Who should I care about? Who is my neighbor? What should the neighbors or passers-by have done in these cases? What would each of us do in a similar situation? Do I have a moral obligation to try to help unfortunate victims such as Ms. Genovese or Mr. Waters or gypsies? What does it mean to be a moral person in the world? Why should I be a moral person? What is morality anyway? Why do we even need morality? Is it in my interest to be a moral person? What is the basis of morality? What does it mean to be moral when it involves personal sacrifice? Such questions serve as an entry into an aspect of philosophy which deals with the question how should we live. That aspect, moral philosophy, is also referred to as ethics.

6.2   ENGINEERING AND CODES OF ETHICS

Engineering is the discipline of acquiring and applying scientific and technical knowledge to the design, analysis, and/or construction of works for practical purposes. ABET defines engineering as: "The creative application of scientific principles to design or develop structures, machines, apparatus, or manufacturing processes, or works utilizing them singly or in combination; or to construct or operate the same with full cognizance of their design; or to forecast their behavior under specific operating conditions; all as respects an intended function, economics of operation and safety to life and property" (Holmes, 2002). The broad discipline of engineering encompasses a range of specialized sub-disciplines that focus on the issues associated with developing a specific kind of product, or using a specific type of technology.

Engineering, similarly to law and medicine, is a profession, which means that it is an occupation, vocation or career where specialized knowledge of a subject, field, or science is applied (Wikipedia, 2008). It is usually applied to occupations that involve prolonged academic training and a formal qualification. It is axiomatic that "professional activity involves systematic knowledge and proficiency" (Kashner, 2005). Professional bodies that may set examinations of competence, act as a licensing authority for practitioners, and enforce adherence to an ethical code of practice usually regulate professions.

Engineering ethics is the field of applied ethics, which examines and sets standards for engineers' obligations to the public, their clients, employers and the profession (Petroski, 1985). Engineering does not have a single uniform system, or standard, of ethical conduct across the entire profession. Ethical approaches vary somewhat by discipline and jurisdiction, but are most influenced by whether the engineers are independently providing professional services to clients, or the public if employed in government service; or if they are employees of an enterprise creating products for sale.

Engineers decide to become members of the profession of their own free will; thus, one key aspect of the codes of ethics or conduct as described throughout the various sub-disciplines of engineering is that they have been set forth and accepted voluntarily as explained by Davis (1999). A second point offered by Davis is that ethical codes by their very nature are moral. One cannot speak of immoral ethical codes or principles; there is no such phenomenon as a Nazi code of ethics.

Engineering ethics is an example of the category in ethics referred as applied ethics. Applied ethics, which typically deals with controversial moral problems, is one of the key divisions within the study of ethics, which also includes descriptive morality (i.e., the actual beliefs, customs, principles and practices of people and cultures) and moral philosophy or ethical theory (i.e., the systematic effort to understand moral concepts and justify moral principles). A legitimate question one might ask is why we as engineers should be concerned with moral philosophy? Is not it enough to simply know and embrace our codes of ethics as engineers? Perhaps, but we would argue that such may not be the case at all. As argued by Pojman and Fieser (2009), "Theory without application is sterile and useless, but action without a theoretical perspective is blind." In today's ever shrinking planet, the need for an awareness of, an appreciation for and, ideally, an understanding of the diverse beliefs, customs, and principles of the various cultures and societies throughout the world appears to be more important today than ever before. As the inexorable forces of globalism bring us even closer, that need is going to become even more important in the future.

6.3   ETHICS AND THE ETHICAL SEQUENCE

Fundamentally, an ethic from a philosophical perspective is a differentiation of social from anti-social behavior. From the perspective of a systems engineering, a different kind of perspective for example, an ethic can be seen as a limitation of freedom of action in support of the advancement or health of the system as a whole. Contrast these two views with that of an ecologist who may imagine an ethic as limitation on freedom of action in the struggle for existence. Whatever the discipline, an ethic has at its origin the tendency for interdependent individuals or groups to evolve modes of co-operation. A sociologist may refer to the holistic structure as a community; an ecologist may refer to it as an eco-system, and a political scientist may refer to it as a government. And what would be the equivalent for an engineer? Who and what matters for the engineering professional? The answers to such questions are the focus of this present text.

Rather than describing an ethic whether pure or applied, perhaps, it is more appropriate to consider an ethical sequence. Such a sequence would move from the simplest of ethics, which deal with the relationships among individuals to the relations between the individuals and the various communities of which that individual is a part to the interdependency of humankind to the environment. The *Ten Commandments* or *Mosaic Decalogue* (Catholic Encyclopedia, 2008) is an example of an ethic, which focused in part on relationships among individuals[2]. Gandhi sought to transform such restrictions from the negative to a positive construction. In contrast to the *Mosaic Decalogue* of negatives, Gandhi (Singh, 2004) provided seven root causes of unfairness and injustice, all consisting of volitional human activities in the absence of socially redeeming moral content[3].

The integration of an individual into a greater community or society is the subject of the ethic of reciprocity or *The Golden Rule* (Teaching Values, 2008), a fundamental moral principle that simply means, "treat others as you would like to be treated." *The Golden Rule* may be the most basic of all ethics. The rule can be found in various forms in a wide array of the earth's wisdom traditions. According to the Greeks, "What you wish your neighbors to be to you, such be also to them" (Graves,

---

[2] The Ten Commandments according to many scholars serve a dual purpose; they form a covenant between God and his people, and serve as a moral law, or code, by which his people are to live by. To achieve the full observance of the Ten Commandments, an elaborate system, known as Mosaic Law, was put into place. This Mosaic Law involved a vast legal system of Israel, civil, criminal, judicial, and ecclesiastical framed after the Decalogue. The Mosaic Law was to be a temporary experiment, while the Decalogue was to be permanent.

[3] Gandhi's Seven Root Causes include: wealth without work; pleasure without conscience; knowledge without character; commerce without morality; science without humanity; worship without sacrifice; and politics without principles.

[4] Mahatma Gandhi (1869—1948) lead India to independence using non-violent means. Indians refer to him as Mahatma Gandhi which means great soul. Martin Luther King later used these same techniques to fight for black civil rights in the USA.

Figure 6.3. Mahatma Gandhi.[4].

Figure 6.4. The 14th Dalai Lama.

1876). In the modern world, the Dalai Lama has stated: "If you want others to be happy, practice compassion. If you want to be happy, practice compassion"[5].

The notion of a sequence moving towards greater complexity parallels the evolutionary process that characterizes the unfolding of the universe from single particle to immense gas clouds to the formation of stars and planets and ultimately life.

It is not an oversimplification to state that all ethics, at least those developed so far, have as their foundation the proposition that the individual is a member of a community of interdependent parts. The identification of the various members of the community and their rejection or inclusion into the dialog may yield very different answers when confronted with an ethical dilemma whether it is in engineering, some other discipline or life itself.

Challenge Box: Were you surprised that all of the major world wisdom traditions include some form of the Golden Rule? Explain. What are the implications for you?

Do you believe our sense of ethical responsibility is evolving? Will it reach a steady state? What would such a steady state look like? Or will it continue to grow inexorably? What are the implications for you as you begin your career?

---

[5] His Holiness the XIVth Dalai Lama, Tenzin Gyatso, is the spiritual and temporal leader of the Tibetan people. The Dalai Lamas are the manifestations of the Bodhisattva of Compassion, who chose to reincarnate to serve the people. Dalai Lama means Ocean of Wisdom.

Landmines and the War in Iraq

Figure 7.1. Martin Luther King.

> It is no longer a choice, my friends, between violence and nonviolence. It is either nonviolence or nonexistence. And the alternative to disarmament, the alternative to greater suspension of nuclear tests, the alternative to strengthening the United Nations and thereby disarming the whole world, may well be a civilization plunged into the abyss of annihilation, and our earthly habitat would be transformed into an inferno that even the mind of Dante could not imagine.
>
> Martin Luther King, Jr. (King et al., 1990).

Figure 7.2. George McGovern.

> I'm fed up to the ears with old men dreaming up wars for young man to die in.
>
> George McGovern.

A recent report documents the effects of the 2003 war in Iraq and the ensuing period after the fall of the Baathist regime on health, the health system, and the health system reconstruction. The World Health Organization (WHO) defines health as follows: "a state of complete physical, mental and social well-being and not merely the absence of disease or infirmity." The WHO argues that the impact of war and violence, in addition to the deaths and injuries due to weaponry, also includes the often greater longer-term suffering linked with damage to the essential infrastructure, a poorly functioning health system and the failure of relief and reconstruction efforts (Panch, 2004).

It may be useful to explore briefly what is meant by war. According to Cook (2004)

What, however, is war? According to the eminent Prussian soldier and theorist Von Clausewitz (2007), "War is …an act of violence to compel our opponents to fulfill our will." Clausewitz continues to unpack the role of violence in its relation to will. "Violence, that is to say, physical force (for there is no moral force without the conception of States and Law), is therefore the means; the compulsory submission of the enemy to our will is the ultimate object." Therefore, American warriors are, as Vice Admiral James B. Stockdale (2004) said, "in the business of breaking people's wills and the most important weapon in breaking people's wills is not the fire power but to hold the moral high ground." According to Clausewitz, Stockdale was convinced, that "War was not an activity governed by scientific laws, but a clash of wills, of moral forces." To make the same point in another way, Napoleon's succinct words are much more vivid: "In war, the moral is to the physical as three to one" (Asprey, 2002).

The direct impact of war/conflict on health is set forth by the WHO and shown in Table 7.1.

Table 7.1: The Direct Impact of Conflict on Health	
WHO Report on Violence and Health (Panch, 2004)	
Increased mortality	• Physical trauma • Infectious diseases • Deaths avoidable through health care
Increased morbidity	• Injuries due to physical trauma • Injuries due to increased societal violence • Infectious diseases • Reproductive health issues • Nutrition • Mental health
Increased disability	• Physical • Psychological • Social

7.1   IMPACT OF THE WAR IN IRAQ

An analysis of a nationwide survey of Iraqi households estimates that in excess of 100,000 deaths occurred since the 2003 invasion and possibly many more (Pinch, 2004). Most deaths were the result of air strikes by coalition forces. More than half of those reported killed by coalition forces were women and children. In other words, more than 500,000 women and children have been killed in Iraq since the outset of the 2003 war. No figures are available on civilians injured during the conflict. Typically, it is estimated that the number of people injured is three times the number of deaths though evidence suggests that due to widespread terrorist activities and coalition force responses, the ratio may be as high as ten to one. That equates to likely numbers of civilian injuries ranging from 300,000 to 1,000,000 with more than half likely women and children. Compounding the tragedy of increased civilian injuries is the estimated 10 million landmines and explosive remnants of war in northern Iraq alone which could take up to approximately 15 years to clear (Pacific Disaster, 2004).

The impact of the conflict on both the health system and the health sustaining infrastructure has also been profound. With respect to various health services, there has been a marked reduction in security, financial exclusion and geographical exclusion. In Iraq, there is a shift in emphasis from primary care to specialist care, a reduction in rural and community based services and compromised public health programs. The country has seen widespread destruction of health clinics and is characterized by a lack of needed drugs and a lack of proper equipment maintenance. Transport of vaccines has been particularly difficult because of the need for sustained, cold temperatures in route.

7.2   LANDMINES

One of the most controversial weapon systems used in modern warfare is the land mine. There are between 70 and 80 million land mines in the ground in one-third of the world's nations. Landmines are indiscriminate weapons that maim or kill 15,000 to 20,000 civilians every year. They cost as little as $3 to produce, but as much as $1000 to remove.

The presence of land mines threatens people's lives, and also prevents much-needed economic growth and development. Long after wars are over, land mines make land unusable for farming, schools or living, preventing people from rebuilding lives torn apart by conflict.

Land mines are explosive material contained in casings of metal, plastic or wood that detonate from the pressure of a footstep (anti-personnel mine) or a passing vehicle (anti-tank mine). Children are just as likely to step on a land mine as a soldier. A land mine typically includes the following components: firing mechanism or other device (including anti-handling devices); detonator or igniter (sets off the booster charge); booster charge (may be attached to the fuse, or the igniter, or be part of the main charge); main charge (in a container, usually forms the body of the mine); and casing (contains all of the above parts).

Land mines are an ancient invention dating back 600 years. Precursors of conventional landmines appeared in the 15th century at the Battle of Agincourt in England and in the 18th century during the American Civil War. After his troops encountered these devices, the commander of the Union Army, General William T. Sherman, said that the use of landmines "was not war, but murder" (Sherman, 2000).

In the 20th century, land mines were developed to meet new threats including tanks and armored vehicles. More than 300 million anti-tank mines were used during World War II alone. WWII also saw an increase in the use of anti-personnel land mines. Since anti-tank mines could be removed by the enemy, anti-personnel mines were placed around them as guards. One of the most effective anti-personnel types was the German-made "bouncing betty," designed to jump from the ground to hip height when activated, propelling hundreds of steel fragments over a wide range. Noting their effectiveness, militaries began to use anti-personnel mines as weapons in their own right.

Since 1945, land mines have been used throughout the world in wars of liberation, civil wars, and local conflicts, with a devastating impact on economic and political reconstruction. Land mines demoralize survivors as well as their families and communities, impeding the process of peace and reconciliation. Often carefully and cleverly hidden, population movement is restricted, land cannot be cultivated, roads and bridges cannot be rebuilt and refugees cannot return to their homes. Survivors of land mine accidents often cannot work at their previous jobs and require retraining. Without a strong workforce, the pace of reconstruction slows.

For poor countries, clearing land mines places an additional financial resource burden. Typically, it costs from $300 to $1,000 to locate and destroy a single landmine and an additional $100 to $3,000 to provide an artificial limb to a land mine survivor. This cost does not go away as children and adults must replace these devices on a yearly basis.

The international community is making a concerted effort to eradicate land mines with the passage of the Mine Ban Treaty (also known as the Ottawa Convention) which came into force on March 1, 1999. International, nongovernmental and private-sector organizations are working with affected countries to establish mine action campaigns. As of 2008, the United States has refused to sign the treaty. According to Bush Administration, the United States will not join the Ottawa Convention because its terms would have required it to give up a needed military capability (CNN, 2004). This reverses a position previously held by the Clinton Administration (Landmines, 2008).

7.3 CONCLUSIONS AND REFLECTIONS

One of the most active individuals in mounting a public campaign to outlaw the use of land mines world wide is Bianca Jagger[1]. She has dedicated a large portion of her professional life to the issues of landmines and cluster bombs. In a recent article (Jagger, 2007), she described her position concerning the position and policy of the British government:

> "Faced with a growing body of data and expert testimony that these self-destruct mechanisms (of the landmines placed in a war time situation) don't work, UK officials have

---

[1]Bianca Jagger is a social and human rights advocate and a former actress. Jagger is a Council of Europe Goodwill Ambassador, Chair of the World Future Council, Chair of the Bianca Jagger Human Rights Coalition, and a member of the Director's Leadership Council of Amnesty International U.S.

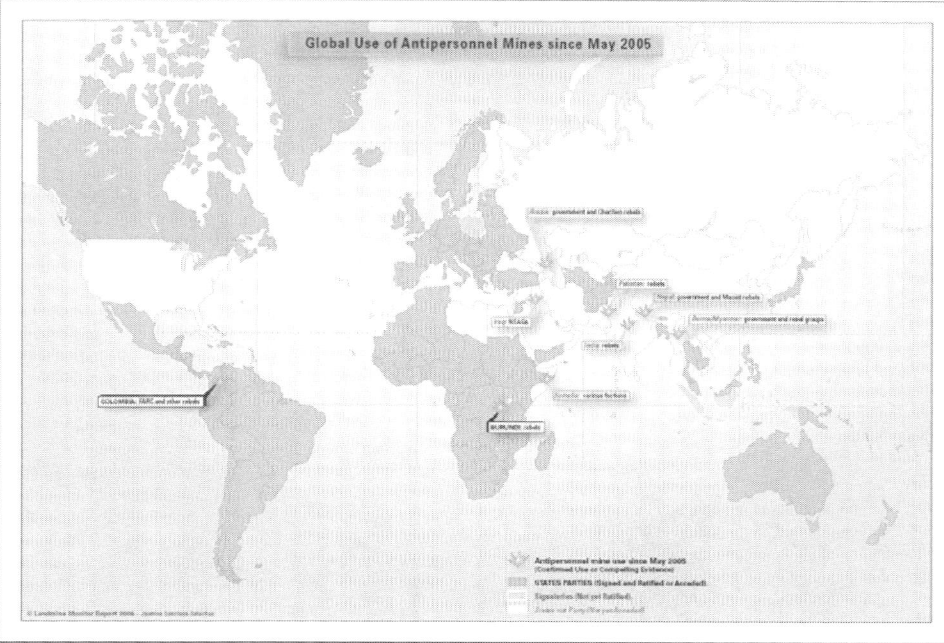

Figure 7.3. Global use of antipersonnel mines since May 2005. (Landmine Monitor Report 2006, International Campaign to Ban Landmines. Reposted with permission, http://www.icbl.org/lm/2006/)

focused on discrediting the evidence not on examining the implications of this for the protection of civilians. Despite the rhetoric, close scrutiny shows that protection of civilians is rarely at the forefront of Government thinking on this issue. In March this year the Government announced that it had completed a review that had 'considered carefully the humanitarian factors' associated with cluster bombs. Yet officials have refused to reveal what evidence regarding civilian harm was actually considered - it is known that the assessment contained no statistics regarding civilian casualties, nor on the quantities of unexploded bombs left in different post-conflict environments, nor statistics on the areas of land painstakingly cleared by civilian teams and funded out of precious development monies."

Jagger added:

"It seems to be based on two premises - one which is misguided and the other which is simply callous. The first is a lazy assumption that protecting civilians means putting British troops at risk. Yet serious military thinkers recognize that this is not the case: in

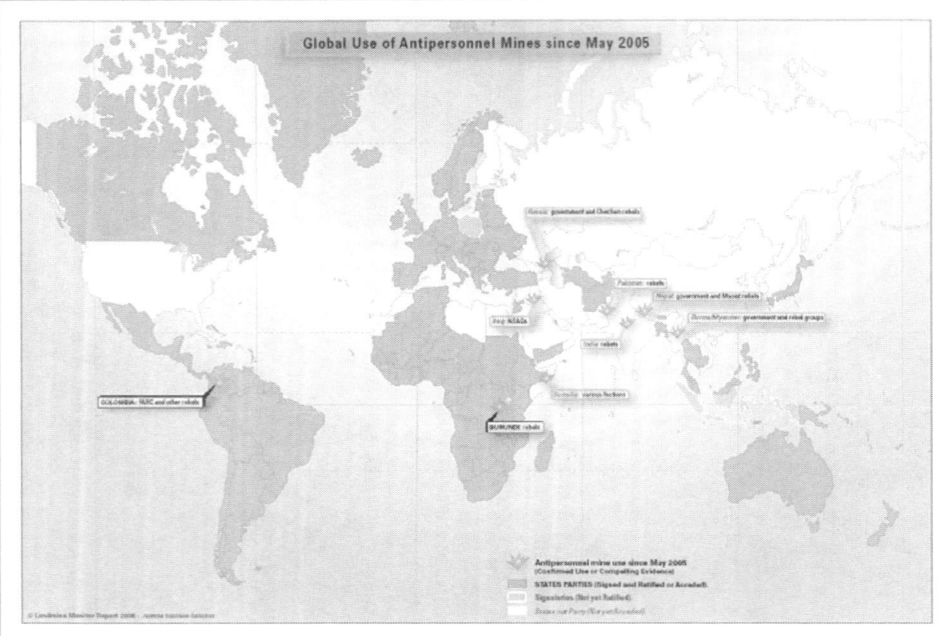

Figure 7.4. 1997 Convention on the prohibition of the use, stockpiling, production, and transfer of antipersonnel mines and their destruction (Landmine Monitor Report 2006, International Campaign to Ban Landmines. Reposted with permission, `http://www.icbl.org/lm/2006/`)

modern war, killing civilians fuels the enemy. The second premise is that foreign civilians do not really count. Whilst robust methods are used to protect domestic populations against anything from dangerous pharmaceuticals to faulty toys, there is no equivalent diligence when the likely victims are foreign, disempowered and voiceless."

It is the last comment that warrants further reflection. Killing civilians adds to the intensity of wars and resoluteness of the population to continue the struggle no matter the devastation. This was certainly the case in World War II for both the Germans and the Japanese. Perhaps even more disturbing is the observation that while we spend great sums and invest considerable time and effort in careful consideration of the safety of toys for our children and drugs for our population, there is virtually no consideration given to victims who may be foreign, or disempowered and or voiceless.

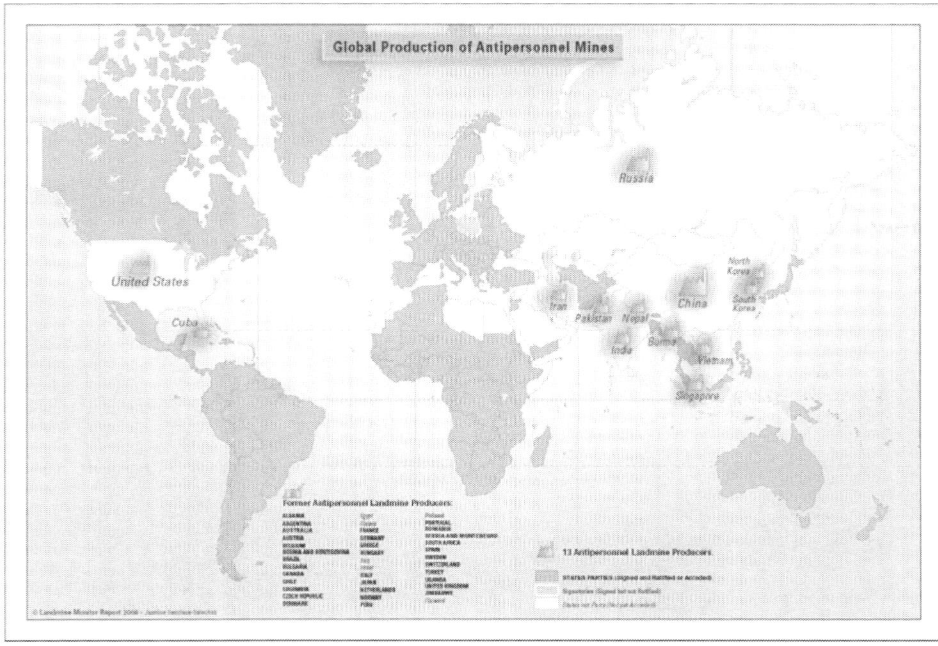

Figure 7.5. Global production of antipersonnel mines (Landmine Monitor Report 2006, International Campaign to Ban Landmines. Reposted with permission,
http://www.icbl.org/lm/2006/)

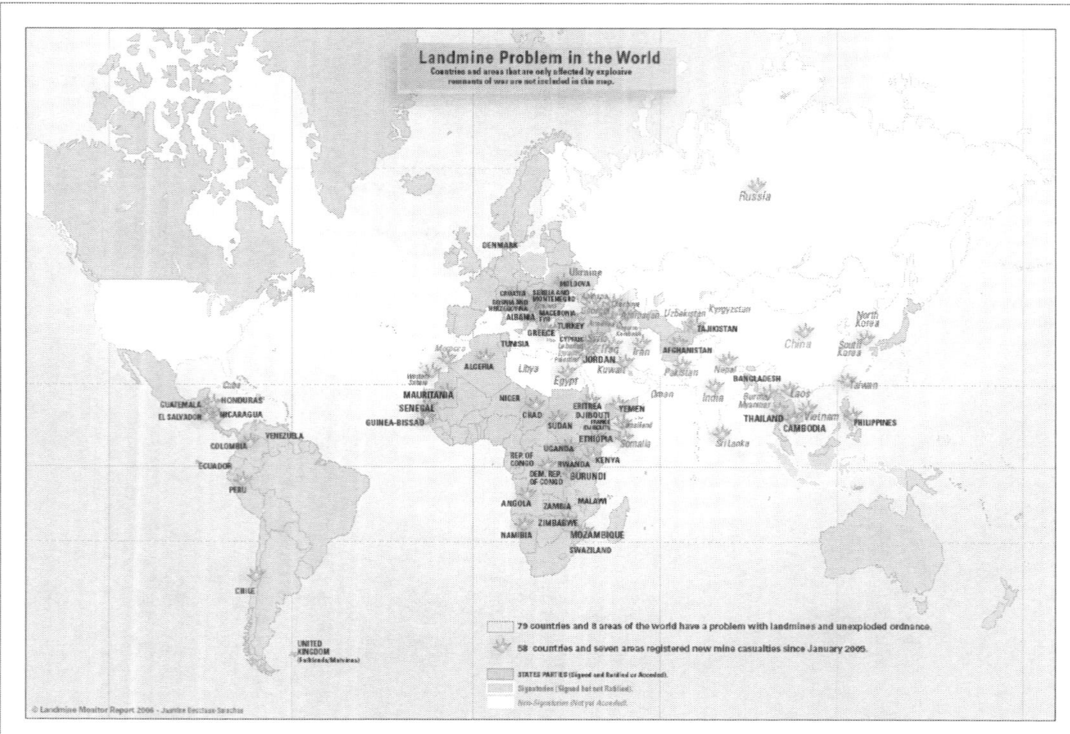

Figure 7.6. Landmine problem in the world (Landmine Monitor Report 2006, International Campaign to Ban Landmines. Reposted with permission, http://www.icbl.org/lm/2006/)

Challenge Box: What do you personally know about the impact of war? Is your information from first hand sources or from the media? Can war ever be justified, that is, what is a just war?

Many engineers work in war-related industries. We have designed the weapon systems. Universities seek out defense related funding aggressively to support both graduate and undergraduate programs. Do we in engineering have any special sense of ethical responsibility for war? And what of the engineering curriculum, do you feel that adequate attention is paid to the issues related to conflict and war in your studies? Do you believe the issues should even be included or are those more appropriate for other fields of study?

Consider the following scenario: You finish up your formal training and get a spectacular offer from an engineering firm that consults world-wide in a wide array of topics. The firm is just awarded a new contract to design the next generation of land mines for use in protecting the interests of the U.S. abroad. Because of your expertise, you have been chosen project director...a guaranteed path to advancement. What are you thoughts and feelings concerning the possible position?

C H A P T E R   8

Hurricane Katrina and the Flooding of New Orleans

Led Zeppelin, *When the Levee Breaks* (News Hounds, 2005).

If it keeps on rainin' levee's goin' to break
If it keeps on rainin' levee's goin' to break
When The Levee Breaks I'll have no place to stay.
Mean old levee taught me to weep and moan
Lord mean old levee taught me to weep and moan
Got what it takes to make a mountain man leave his home
Oh well oh well oh well.
Don't it make you feel bad
When you're tryin' to find your way home
You don't know which way to go?
If you're goin' down South
They go no work to do,
If you don't know about Chicago.
Cryin' won't help you, prayin' won't do you no good,
Now, cryin' won't help you, prayin' won't do you no good,
When the levee breaks, mama, you got to move.
All last night sat on the levee and moaned,
All last night sat on the levee and moaned,
Thinkin' 'bout me baby and my happy home.
Going, go'n' to Chicago,
Go'n' to Chicago,
Sorry but I can't take you.
Going down, going down now, going down.

Randy Newman, *Louisiana 1927* (Common Dream).

What has happened down here is the wind have changed
Clouds roll in from the north and it started to rain
Rained real hard and rained for a real long time
Six feet of water in the streets of Evangeline
The river rose all day
The river rose all night
Some people got lost in the flood
Some people got away all right
The river have busted through clear down to Plaquemines
Six feet of water in the streets of Evangeline
CHORUS
Louisiana, Louisiana
They're tryin' to wash us away
They're tryin' to wash us away
Louisiana, Louisiana
They're tryin' to wash us away
They're tryin' to wash us away

Over the course of the last 100 years, two floods have devastated the city of New Orleans and surrounding areas: the Great Flood of 1927 and the aftermath of Hurricane Katrina in 2005. Both events have been immortalized by artists with two examples in music provided. There are many others.

8.1   THE GREAT FLOOD OF 1927

As the 1927 flood approached New Orleans, about 30 tons of dynamite were set off on the levee at Cameron, Louisiana and sent 7,000 m^3/s of water pouring through. This prevented New Orleans from experiencing serious damage, but flooded much of St. Bernard Parish and all of Plaquemines Parish's east bank. As it turned out, the destruction of the Caernarvon levee was unnecessary; several major levee breaks well upstream of New Orleans, including one the day after the demolitions, made it impossible for flood waters to seriously threaten the city. There is some belief that the purpose of the levee explosion was to save the wealthier parts of the city by directing the flow of water to the more rural, less developed communities in order to minimize financial losses.

Figure 8.1. Breach of levee, 1927.

Figure 8.2. Flooding in Morgan City, 1927.

Figure 8.3. Barge, near Lake Pontchatrain, 1927.

8.2   HURRICANE KATRINA

Seventy five years later an even more devastating flood would create great tragedy again for the residents of New Orleans. On August 29th, 2005, the nation watched, in horror, video images of New Orleans sliding into utter chaos. Initially, there was some hope that New Orleans had been spared the full effect of the long-predicted Category 4 hurricane[1], Katrina. This initial hope was soon replaced by horror and disbelief as two levees designed and built to hold back the waters of Lake Pontchatrain failed and the city of New Orleans began to fill with water.

Figure 8.4. NOAA satellite image of hurricane Katrina (NOAA, 2005).

The extent of the tragedy remains indelibly engraved in the minds of anyone who remembers those images. Water filled in places to the rooftops, almost immediately, in the Lower Ninth Ward, home to some of the poorest people in the area, St. Bernard Parish and New Orleans East. People, who had not been able to follow, believe or even hear the evacuation orders, were trapped in their

[1] Hurricane strengths are judged according to the Saffir-Simpson Hurricane Scale is a scale classifying most Western hemisphere tropical cyclones that exceed the intensities of tropical depressions and storms, and thereby become hurricanes. The categories into which it divides hurricanes are distinguished by the intensities of their respective sustained winds. The classifications are intended primarily for use in measuring the potential damage and flooding a hurricane will cause upon landfall.

homes, forced to climb to the second floors and then to the attics, finally to break though their own roofs with nothing more than a pocket knife where they waved desperately for someone to rescue them. Others waited on sidewalks, luggage packed, for the buses they thought were coming to take them to safety, eventually giving up and returning to their houses where they were trapped by the rising water.

In the aftermath of Hurricane Katrina, engineers investigated the possibility that a failure in the design, construction, or maintenance caused much of the flooding. Originally, it was speculated that the levees had been overtopped by the storm-surge; however, this was later found not to be the case. Some investigations pointed to the possibility of a weakening of the soil beneath the foundations of the flood walls due to storm water caused the ground to shift, which would indicate that a major design flaw made during the construction of the levees had been a major cause of the failures due to the storm.

Furthermore, the region's natural defenses, the surrounding wetlands and barrier islands have been dwindling in recent years. Much of the land was undeveloped wetlands on the lake side, and only small levees were constructed in the 19th century. A much larger project to build up levees along the lake and extend the shoreline out by dredging began in 1927. As the city grew, there was increased pressure to urbanize lower areas, and, as a result, a large system of canals and pumps was constructed to drain the city. Drainage of the formerly swampy ground allowed more room for the city to expand, but also resulted in subsidence of the local soil.

Outside of the city, the Mississippi River's natural deposition of suspended sediment built up the river's delta marshlands during periodic flooding episodes. However, the lower Mississippi was later restricted to channels for the benefit of shipping, which interrupted the process that continued to build the delta region south and east from New Orleans and prevented its erosion. As the swampy lands of Southern Louisiana shrank, the land began to sink. Entire barrier islands disappeared during periodic storms as the land of the vast delta slowly settled without river silt to replenish the wetlands. Approximately one-third of the land subsidence has been attributed to the large number of canals through the delta. Barge traffic and tides erode the earth around the edge of the canals, and salty water from the Gulf of Mexico water seeps in along them, slowly salinating the ground and killing the vegetation that the land previously depended on to anchor it.

The flooding of metro New Orleans was an avoidable man-made disaster. The levee and canal walls failed because of human errors. "Experts say the New Orleans flood of 2005 should join the space shuttle explosions and the sinking of the Titanic on history's list of ill-fated disasters attributable to human mistakes" (Gelinas, 2007).

The U.S. Army Corps of Engineers, a federal agency has sole authority over the design and construction of metro New Orleans' flood protection and water management as authorized by Congress in the Lake Pontchartrain Hurricane Protection Project. The U.S. Army Corps of Engineers has accepted that faulty design specifications and substandard construction of certain levee segments, not a hurricane were the primary cause of the flooding damage in the New Orleans area.

Figure 8.5. Aerial view of the flooding in part of the central business district (The Superdome is at center.) (Access Kansas, 2005).

Figure 8.6. Flooded I-10 interchange and surrounding area of northwest New Orleans and Metairie, Louisiana (Wikipedia, 2008).

Today, almost three years have passed. Throughout much of New Orleans, recovery from Katrina has been hindered by the city's many pre-storm weaknesses and delayed by false starts. Entire city blocks of the Lower 9th Ward closest to the Industrial Canal breach are now a field of prairie grass pockmarked with concrete-slab foundations and driveways ending in wildflowers. Community anchors are boarded up or abandoned, scarred by fire and rotting from water. The federally funded Road Home program, the largest housing recovery program in U.S. history, has given out $6.4 billion to storm victims looking to sell or rebuild their homes. There seems to be a growing momentum for restoration as well as something much more important, hope.

> **Challenge Box:** Should we care about New Orleans? Why or why not? As engineers, do we have a special responsibility for the Katrina-caused damage to the city and the loss of life? Should we help New Orleans rebuild? Should the residents of the city be allowed to return?

Unfortunately, there's no escaping the blocks of warped and ruined homes. Pontchartrain Park, north of Gentilly and bordering the lake, was occupied primarily by retirees. Most residents who lived near Lake Pontchatrain were overwhelmed by the tragedy and the needed effort to come back, and as a result, areas such as Pontchatrian Park and Gentilly remain ghost towns. The city-wide population of New Orleans is far below its pre-Katrina level of 454,000. The numbers are in dispute, most reliable estimates are around 308,000—a rebound of about 180,000 since the months after the storm. Most agree resettlement is leveling off. New homes are being built higher and "smarter," but the emphasis for flood protection remains with the levees. While the Army Corps of Engineers argues that dramatic repairs have significantly strengthened the system, critics like Robert Bea, a civil engineering professor at the University of California-Berkeley, are dismissive. "It may be able to withstand a Category 3 storm," he says, "but not likely." That level of protection won't be fully in place until 2011. And it's unlikely the system will ever be strong enough to protect against a Category 5 storm."

> **Challenge Box:** The levee system was an engineering failure, yet all specifications/codes were met. What does that say to you about the notion of safety? Is there a responsibility beyond the regulations, the codes and the laws?
> Many of the citizens of New Orleans who could not escape the city before the storm hit were the poor or elderly, children or the infirmed. Is there any additional responsibility for these individuals? Why or why not?

More than 150,000 jobs were lost in the storm. The city has had difficulty even ascertaining who owns what property, especially in the Lower Ninth Ward, where many houses passed through generations of families with no changes in the deeds. And at least half the city's pre- storm residents were renters, few of them with insurance. Now the city is aggressively pushing home ownership for all.

CHAPTER 9

Disappearing Bumble Bees

Figure 9.1. E.O. Wilson.

> *We should preserve every scrap of biodiversity as priceless while we learn to use it and come to understand what it means to humanity.*
>
> *E.O. Wilson (Brainy Quote, 2008).*
>
> *One way to think about the sun, every time you see it at dawn, is to think of it as an act of cosmic generosity.*
>
> *Brian Swimme (Wikipedia, 2008).*

Figure 9.2. Emerson.

> *Nature is a mutable cloud which is always and never the same.*
>
> *Ralph Waldo Emerson (Quoteworld, 2008).*

In the U.S., bees are responsible for pollinating up to one-third of the food supply, yet they have been dying off in alarming numbers for almost two decades. Faced with such detrimental foes as parasitic mites, pesticides, and urban development, bees are at a formidable disadvantage in our modern world. But since crops are so dependent on the work of these tiny creatures, beekeeping has become a big business, and maintaining effective hives is practically a competitive sport. Honeybees, even as they dwindle in number, are the most dependable source of pollination, carrying much of the burden to sustain our agricultural needs.

Figure 9.3. Honey bees serve as pollinators.

9.1   DEVELOPING CRISIS

Honey bees pollinate more than 90 cultivated crops, including avocadoes, cucumbers, watermelons, citrus fruit and, notably, almonds; California's almond industry alone needs about half the country's 2.5 million commercial hives for pollination every year. Honey bees are responsible for more than $20 billion in annual pollination value and one-third of the food we eat, from vegetables to oils to meat, from animals that graze on pollinated foliage. Managed honey bees are not the only pollinators in the air as, in the Pacific Northwest alone, there are some 900 species of native bees. Some plants, such as tomatoes, rely on wind pollination. But because modern agriculture demands high yields from densely planted crops, they need modern commercial honey bees in massive quantities.

Wild bees and the flowers they pollinate are disappearing together in Britain and the Netherlands, researchers have announced. It is not clear which started to disappear first, the bees or the flowers, but the trend could affect both crops and wild species, the researchers report in a recent issue of the journal *ScienceDaily* (2007).

An undiagnosed honeybee ailment spread across the Northern Hemisphere has left British beekeepers worrying about how many of their bees survived the winter. The implications for agricul-

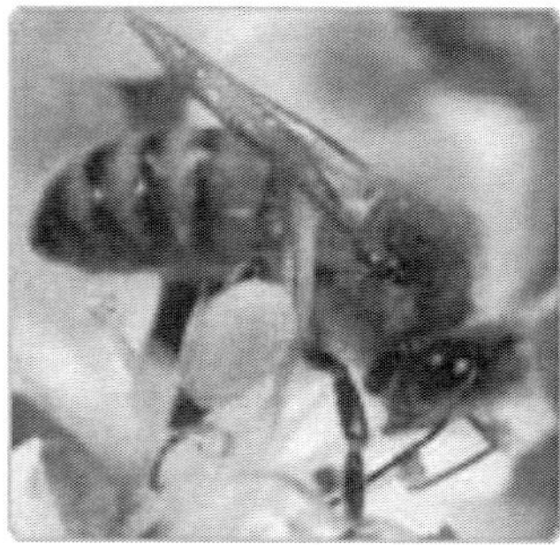

Figure 9.4. Honey bees in marked and mysterious decline.

tural pollination and production are huge. No one knows the cause of the honeybee sickness, which caused the deaths of thousands of honeybee colonies across the US and Europe.

This follows a series of unexplained, but very severe, honeybee colony losses over the past few years in Poland, Greece, Italy, Spain and Portugal and heavy losses in other countries are suspected to be going unreported. The symptoms of the colony deaths are varying across Europe and North America and the losses generally come to light between late summer and early spring. This winter in the USA, colonies have dwindled as the older bees have died leaving behind the queen and young workers not yet ready to forage for pollen and nectar and insufficient in number to maintain the colony. In the United Kingdom this past year, there were a few but significant examples of what became termed the Marie Celeste phenomenon - colonies simply disappearing from hives leaving no bees for post-mortem analysis.

Challenge Box: Should we care about rapid decline of bumble bees? Why or why not? Do bees matter as pollinators for our food or do they matter in their own right? Do we need to worry more generally about the loss of wildlife? What can we do about this decline? Is there a connection between the plight of the bumble bees and engineering? If it exists, what is that connection?

In the closing months of 2006, thousands of American bee hives were found to be almost entirely devoid of bees, victims of a mysterious phenomenon now known as Colony Collapse Disorder (CCD). A study of 150,000 managed bee colonies in 15 states, commissioned by the Apiary

Inspectors of America, found that from September 2006 to March 2007, roughly one-third of the colonies were lost.

Bee keepers have suffered similar unexplained losses in the past, and not all of the hives in the survey were lost to whatever is causing colony collapse. But people are understandably worried that the disorder may threaten all three million managed bee colonies in the United States, a $14.6 billion commercial pollination business. As a result, it has become urgent that scientists determine what is causing the colonies to disappear and how many more colonies stand to vanish.

9.2   COLONY COLLAPSE DISORDER MODEL

Many scientists have suggested that some kind of virus or bacterium — or some combination of infectious agents, possibly carried by parasites like mites — is killing the bees. A model was developed and examined in 2003 to determine whether severe acute respiratory syndrome could be controlled (Ellis, 2009). A similar model may be used to study CCD.

A colony which has collapsed from CCD is generally characterized by all of these conditions occurring simultaneously: (a) a complete absence of adult bees in colonies, with little or no build-up of dead bees in or around the colonies; (b) a presence of capped brood in colonies. Bees normally will not abandon a hive until the capped brood have all hatched; (c) Presence of food stores, both honey and bee pollen: which are not immediately robbed by other bees and which when attacked by hive pests, the attack is noticeably delayed. Precursor symptoms that may arise before the final colony collapse are: (a) an insufficient workforce to maintain the brood that is present; (b) a workforce made up of young adult bees; (c) a queen is present; and (d) colony members are reluctant to consume provided feed.

> **Challenge Box:** How do you react to the assertion that bio-engineering is either a futile exercise or an ecological disaster with no possibility for recall? Explain your position.
> Explain in your own words what is meant by a mechanical universe mindset. Do you have one? Give examples which reinforce your assertion.

Perhaps the disappearance of the bee colonies point to far broader issues that need to be considered such as the diminishing biodiversity in our planet's ecosystem, the widespread use of pesticides and herbicides, unintended gene transfer from genetically modified crops, direct and indirect stress on the ecosystem, global climate change, navigational hindrances and, in our opinion the most significant, a "mechanical universe" mindset. Such a mindset reduces an infinitely complicated world of interactions to an overly simplistic viewpoint. In mathematics $1 + 1 = 2$, in biology, $1 + 1$ may equal 3, or a billion and three. Some argue that the term bio-engineering itself is a contradiction in terms. 'Bio' equates to 'life.' 'Engineering,' as we have seen earlier in this text, refers to something entirely different. Biological forms can never be 'engineered' - i.e., predictably controlled or manipulated. Unlike a sheet of metal that can be machined with consistent results, organisms in natural systems are ever changing and adjusting. This makes 'bio-engineering,' in the best-case scenario, a futile exercise and an enormous misallocation of human and environmental resources, and, in the worse case scenario, an ecological catastrophe with no chance for a product recall.

C H A P T E R   10

Engineering and Traditional Approaches

> *A man is truly ethical only when be obeys the compulsion to help all life which he is able to assist, and shrinks from injuring anything that lives.*
>
> *Albert Schweitzer.*

Figure 10.1. Albert Schweitzer.

As stated described, we shall begin our exploration of making decisions using approaches that are very commonly utilized in engineering today: utilitarianism, rights of persons and virtue ethics. An *Utilitarian* approach emphasizes bringing about the most good that we can. A *rights of person* approach states that actions or decisions are right if each person or "moral agent" is afforded equal respect and consideration. An approach based on *virtue* is identified as the one that emphasizes the virtues, or moral character, in contrast to the approach which emphasizes duties or rules or that which emphasizes the consequences of actions.

10.1 TRADITIONAL APPROACHES TO ENGINEERING ETHICS

10.1.1 UTILITARIANISM

According to Hinman (2008), the basic insights offered by Utilitarianism can be listed as follows: the purpose of morality is to make the world a better place; morality is about producing good consequences not having good intentions; and we should always choose based on bringing the

> *How is one to live a moral and compassionate existence when one is fully aware of the blood, the horror inherent in life, when one finds darkness not only in one's culture but within oneself? If there is a stage at which an individual life becomes truly adult, it must be when one grasps the irony in its unfolding and accepts responsibility for a life lived in the midst of such paradox. One must live in the middle of contradiction, because if all contradiction were eliminated at once life would collapse. There are simply no answers to some of the great pressing questions. You continue to live them out, making your life a worthy expression of leaning into the light.*
>
> *Barry Lopez.*

Figure 10.2. Barry Lopez.

greatest good to the greatest number of people. Harris et al. (2008) offer the following standard for utilitarianism: those individual actions or rules that produce the greatest total amount of utility to those affected are right. Here the most common definition of utility is happiness. utilitarians are in agreement that for most people to be able to pursue utility or happiness, two conditions must be present: freedom and well-being. Freedom is defined as the power to exercise choice and make decisions without constraint from within or without; autonomy; self-determination while well-being amounts to how well a person's life is going for him or her. It may include health, material well-being, food, shelter and education or training. Three approaches to Utilitarianism are the cost/benefit approach, the act utilitarian approach and the rule utilitarian approach.

Harris et al. (2008) suggest the following three steps for implementing a cost/benefit analysis approach to making engineering decisions:

- Assess the available options

- Assess in monetary terms the costs and benefits for all of those affected

- Make the decision that is most likely to result in the greatest benefit for the least cost

Cost/benefit analysis (CBA) is a generic term for a variety of techniques designed to allow decision-makers to determine in a rigorous way whether the payback from a program will be greater than the costs of implementing it. For example, if costs of an environmental program are greater than environmental benefits produced by a program, the program should be abandoned. The economic justification for this use of CBA is the notion that society must decide how to spend its scarce resources and it should spend its money in the most efficient way possible. If money is spent by society on environmental protection programs that do not produce an environmental payback that is greater in economic value than the cost of the program, it is a bad investment and should not be supported. According to CBA theory, public money should be spent on programs that will produce the largest aggregate benefits.

Act Utilitarianism is based on the notion that it is the value of the consequences of *the particular act* that counts when determining whether the act is right. Act Utilitarianism makes no appeals to general rules, but instead demands that the person(s) making the decision evaluate individual circumstances. Harris et al. (2008) offer the following procedure for implementation of an Act Utilitarian approach:

- Determine the possible option available

- Determine the appropriate audience for each of the options

- Remember the principle of *universalizability* must be met and remain consistent. (Immanuel Kant used this term when discussing the maxims, or subjective rules, that guide our actions. A maxim is *universalizable* if it can consistently be willed as a law that everyone ought to obey. The only maxims which are morally good are those which can be universalized. The test of universalizability ensures that everyone has the same moral obligations in morally similar situations

- Select the option that maximizes the good for the previously determined relevant audience

The third utilitarian approach is referred to as Rule Utilitarianism. "Instead of looking at the consequences of *a particular act*, Rule Utilitarianism determines the rightness of an act by a different method. First, the best rule of conduct is found. This is done by finding the value of the consequences of *following a particular rule*. The rule the following of which has the best overall consequences is the best rule. Among early proponents were Austin (1995) and Mills (2008). Rule Utilitarianism is an option for those who believe that there are absolute prohibitions on certain types of actions but do not want to give up on utilitarianism completely. According to Rule Utilitarianism, the principle of utility is a guide for choosing rules, not individual acts.

Challenge Box: Compare and contrast a cost benefit analysis approach, an Act Utilitarian approach an a Rule Utilitarian approach to the following case:

A trolley is running out of control down a track. In its path are 5 people who have been tied to the track by a mad philosopher. Fortunately, you can flip a switch which will lead the trolley down a different track to safety. Unfortunately, there is a single person tied to that track. Should you flip the switch?

As before, a trolley is hurtling down a track towards five people. You are on a bridge under which it will pass, and you can stop it by dropping a heavy weight in front of it. As it happens, there is a very fat man next to you - your only way to stop the trolley is to push him over the bridge and onto the track, killing him to save five. Should you proceed?

10.1.2 RESPECT FOR PERSONS

According to Harris et al. (2008), the moral standard of the ethics of *respect for persons* is straightforward: "Those actions or rules are right that, if followed, would accord equal respect to each person as a moral agent." A moral agent is a person with a capacity for making moral judgments and taking actions that comport with morality. A moral agent is a being capable of those actions that have a moral quality, and which can properly be denominated good or evil in a moral sense. Most philosophers tend to view morality as a transaction among rational parties, i.e., among moral agents, and thus, would exclude animals as well as land and ecosystems from moral consideration. Others (Singer, 2005) state that one must draw a distinction between moral agency and being subject to moral considerations and have argued that the key to inclusion in the moral community is not rationality — for if it were, we might have to exclude some disabled people and infants, and might also have to distinguish between the degrees of rationality of healthy adults — but that the real object of moral action is the avoidance of suffering.

One *respect for persons* approach with which we are very familiar is the Golden Rule, a variant of which appears in the religious and ethical writings of societies and cultures throughout the world (Table 10.1). The Golden Rule is best interpreted as saying: "Treat others only in ways that you're willing to be treated in the same exact situation." To apply it, you are asked to imagine yourself in the exact place of the other person on the receiving end of the action. If you act in a given way toward another, and yet are unwilling to be treated that way in the same circumstances, then you violate the rule.

To apply the Golden Rule appropriately, we need knowledge and imagination. We need to *know* what effect our actions have on the lives of others. In addition, we need to be able to *imagine* ourselves, vividly and accurately, in the other person's place on the receiving end of the action. Application of the Golden Rule can take the following steps:

- Determine the effects on others lives that will result from our action(s).

- Imagine ourselves in the other persons place, that is, 'walk a mile in that person's shoes.'

- Ask ourselves, would we be willing to accept the consequences of our action(s)?

Table 10.1: Comparison of Golden Rule for various cultures and religions.

Culture/Religion	Statement of the Golden Rule
Buddhism	*Hurt not others in ways that you yourself would find hurtful.* Udana-Varga 5,1
Christianity	*All things whatsoever ye would that men should do to you, do ye so to them; for this is the law and the prophets.* Matthew 7:1
Confucianism	*Do not do to others what you would not like yourself. Then there will be no resentment against you, either in the family or in the state.* Analects 12:2
Hinduism	*This is the sum of duty; do naught onto others what you would not have them do unto you.* Mahabharata 5,1517
Islam	*No one of you is a believer until he desires for his brother that which he desires for himself.* Sunnah
Jewish	*What is hateful to you, do not do to your fellowman. This is the entire Law; all the rest is commentary.* Talmud, Shabbat 3id
Taoism	*Regard your neighbor's gain as your gain, and your neighbor's loss as your own loss.* Tai Shang Kan Yin P'ien
Zoroastrianism	*That nature alone is good which refrains from doing another whatsoever is not good for itself.* Dadisten-I-dinik, 94,5

A second *respect for persons* approach is referred to as the Rights Based Approach. According to the Stanford Encyclopedia of Philosophy (2008). "Rights are entitlements (not) to perform certain actions or be in certain states, or entitlements that others (not) perform certain actions or be in certain states. Rights dominate most modern understandings of what actions are proper and which institutions are just. Rights structure the forms of our governments, the contents of our laws, and the shape of morality as we perceive it. To accept a set of rights is to approve a distribution of freedom and authority, and so to endorse a certain view of what may, must, and must not be done." Other philosophers and ethicists suggest that the ethical action is the one that best protects and respects the moral rights of those affected. This approach starts from the belief that humans have a dignity based on their human nature per se or on their ability to choose freely what they do with their lives. On the basis of such dignity, they have a right to be treated as ends and not merely as means to other ends. The list of moral rights -including the rights to make one's own choices about what kind of life to lead, to be told the truth, not to be injured, to a degree of privacy, and so on-is widely debated; some now argue that non-humans have rights, too. Rights imply duties, in particular, the duty to respect others' rights.

Challenge Box: Consider the same two scenarios set forth in the trolley problem but this time applying the Golden Rule first and then the Rights Based approach.

10.1.3 VIRTUE ETHICS

According to its etymology the word virtue (Latin *virtus*) can be interpreted to mean 'possessing courage.' Virtues may be divided into intellectual, moral, and theological. As we are focusing upon engineering and making decisions, we shall limit our discussion to moral virtues only. One particular list of moral virtues includes the following elements:

- Justice: Justice regulates man in relations with his fellow-men. It disposes us to respect the rights of others, to give each man his due. Justice, a condition there of, is the ideal state of humanity, a morally correct state of things and persons. John Rawls[1] claims "Justice is the first virtue of community institution, as truth is of systems of thought." Today many people believe that justice must not be limited or restrained. Many social and political movements are centered, in fact, on the premise of global justice.

- Temperance: Temperance represents a certain quality of self-control and discipline. Individual power lies in slowly tempering desires so that they disappear. Justice then lies in tempering the balance of power so that all maintain equal shares.

---

[1]John Rawls (1921-2002) is considered by many to be one of the most important political philosophers of the 20th century. He wrote a series of highly influential articles in the 1950s and '60s that helped refocus Anglo-American moral and political philosophy on substantive problems about what we ought to do.

Figure 10.3. Virtue.

- Fortitude: Fortitude, which implies a certain moral strength and courage, is the virtue by which one meets and sustains dangers and difficulties, even death itself. It refers to that strength or firmness of mind which enables a person to encounter danger with coolness and courage, or to bear pain or adversity without complaint.

The term "virtue ethics" is a relatively recent one. It is an umbrella term that encompasses a number of different theories. Initially, virtue ethics was characterized as a movement rivaling consequentialism and deontology because it focused on the central role of concepts like character and virtue in moral philosophy. Later versions developed fuller accounts of virtue ethics theories. Most virtue ethics theories take their inspiration from Aristotle, although some versions incorporate elements from Plato[2], Aquinas,[3] Hume[4] and Nietzsche.[5]

[2] Plato (c. 427-347 B.C.E) is one of the world's best known and most widely read and studied philosophers. Known as the student of Socrates and the teacher of Aristotle, he wrote in the middle of the fourth century B.C.E.
[3] Saint Thomas Aquinas, (1225 –1274) was a Dominican priest, a philosopher and a theologian in the scholastic tradition. He was the foremost classical proponent of natural theology, and the originator of the Thomistic school of philosophy and theology.

Virtue ethics is concerned with the person's life as a whole, with character and the kind of person you are. The right perspective on an action, therefore, will for virtue ethics be the one which asks about success in achieving the overall goal, rather than success in achieving the immediate target. What matters is what the person's motivation was, and how this relates to his or her developed character and life as a whole; for this is his or her achievement, what he or she has made of his or her life. Success or failure in achieving the immediate target will affect various judgments we make about the action, but if, like the Stoics, we distinguish clearly between the immediate target and the overall aim, it is achieving the latter, not the former, which will make the action a success. Here virtue ethics differs from theories like consequentialism, for which it is the actual results that matter for our evaluation of the agent (Annas, 2008), and agrees with Kantianism (Kant, 1780), for which what matters is the agent's motivation.

Challenge Box: Let's return to the same trolley problem scenarios. Compare and contrast your responses when you apply a virtue ethics based approach versus the other applied ethical approaches you have already employed. What is the same? What is different?

Which ethical approach was the easiest for you to use? Which one proved most troublesome? Explain.

Did any of the approaches yield answers which you found contrary to your own personal values/beliefs? Explain. Did you find that troubling?

10.2   REVIEW OF EXISTING ETHICAL CODES

At the start of the 21st century, there are as many different codes of conduct in engineering as there are engineering disciplines and specialties. One professional society, the National Society for Professional Engineers (NSPE), has offered one general code which is widely employed today in all the disciplines as well as in engineering education. The NSPE Code of Ethics consists of a preamble followed by a listing of fundamental canons and then rules of practice (NSPE, 2008). The very first canon cautions engineers in the fulfillment of their professional duties, to "hold paramount the safety, health and welfare of the public." As a result, the first rule of practice states that engineers shall "hold paramount the safety, health, and welfare of the public." Note that the explicit requirements focus

---

[4]David Hume (1711-1776) is considered by many to be the most important philosopher ever to write in English. Today, philosophers recognize Hume's work as a precursor of contemporary cognitive science, as well as one of the most thoroughgoing exponents of philosophical naturalism.

[5] Frederick Nietzche (1844-1900) was a German scholar, philosopher and critic of culture.

on the public only; though presumably, concern for the natural world is included implicitly, though only, as it affects humankind.

The American Society of Mechanical Engineers (ASME) sets forth a similarly constructed code of ethics with fundamental principles followed by fundamental canons (ASME, 2008). The first principle states that engineers uphold and advance the integrity, honor, and dignity of the engineering profession by using their knowledge and skill for the enhancement of human welfare. The supportive fundamental canon states engineers shall hold paramount the safety, health and welfare of the public in the performance of their professional duties.

The American Society of Civil Engineers (ASCE) does, at least, mention the environment in its code (ASCE, 2008). According to ASCE, engineers uphold and advance the integrity, honor and dignity of the engineering profession by using their knowledge and skill for the enhancement of human welfare and the environment (fundamental principle) and shall hold paramount the safety, health and welfare of the public and shall strive to comply with the principles of sustainable development in the performance of their professional duties (fundamental canon). There is no explanation of what is meant by the enhancement of the environment. In November 1996, the ASCE Board of Direction adopted the following definition of sustainable development: "Sustainable development is the challenge of meeting human needs for natural resources, industrial products, energy, food, transportation, shelter, and effective waste management while conserving and protecting environmental quality and the natural resource base essential for future development."

The Institute of Electrical and Electronics Engineers (IEEE) Code of Ethics states that its members accept responsibility in making engineering decisions consistent with the safety, health and welfare of the public, and to disclose promptly factors that might endanger the public or the environment (IEEE, 2008). Here, an interesting notion of responsibility towards the environment is described. It is not in opposition to the IEEE code to endanger the public or the environment only to not disclose promptly factors that might endanger the public or the environment.

The Institute of Industrial Engineers (IIE) endorses the Canon of Ethics provided by the Accreditation Board for Engineering and Technology (ABET) whose first principle is that engineers uphold and advance the integrity, honor and dignity of the engineering profession by using their knowledge and skill for the enhancement of human welfare and whose first canon is engineers shall hold paramount the safety, health and welfare of the public in the performance of their professional duties (IIE, 2008). ABET is the accrediting body for all engineering and engineering technology programs in the United States and thus has an important impact on the training of tomorrow's engineers and engineering educators.

Members of the American Institute of Chemical Engineers (AIChE) are challenged to uphold and advance the integrity, honor and dignity of the engineering profession by being honest and impartial and serving with fidelity their employers, their clients, and the public; striving to increase the competence and prestige of the engineering profession; and using their knowledge and skill for the enhancement of human welfare (AIChE, 2008). To achieve these goals, AIChE members shall hold paramount the safety, health and welfare of the public and protect the environment in performance of their professional duties. There is neither elaboration on the idea of protecting the environment nor an identification on from whom or what shall it be protected.

Many other engineering disciplines exist, each with their own codes for ethical conduct. As can be seen from this review, a large percentage of the codes do not explicitly identify the environment as an important stakeholder in discussions of the ethics of engineering choices. Equally as troubling, those codes that do mention the environment refer to the idea of enhancing nature or promoting sustainable development, which is based solely upon meeting human needs. A select few number of codes do mention a responsibility to protect the environment but without identifying from whom or from what.

There are many other engineering disciplines at present, each with its own code of conduct or ethics, which describes the responsibilities of the profession. Most focus heavily on the sense of responsibility engineering has towards employers, society in general and towards other professional engineers.

In their totality, the codes of ethics point to a very different conception or understanding of the natural world then our science provides us with now. We are at once removed from membership in the natural world as there is a listing of responsibilities of the engineering profession to humankind, and if it exists at all, a sense of responsibility to the natural world only in so far as it can provide something for us. We are not products of the earth but somehow placed on it with a focused plan of action set in place to tame it, control it, and to transform it into what suits are interests.

Challenge Box: Compare and contrast the various codes of ethics. Which provided insights when confronted with issues of war, poverty and the environment? Which seemed to closest match your own set of values? Which do you think would be easiest to follow? Which one do you think would be most troublesome? Explain your responses.

Did any of the approaches yield answers which you found contrary to your own personal values/beliefs? Explain. Did you find that troubling?

CHAPTER 11

Engineering and Freedom

Figure 11.1. Freedom.

To this point in our explorations of approaches to making decisions in the 21st century, we have examined traditional applied ethics methodologies; many of which date back to antiquity. We shall now explore several modern notions which may aid us as engineers as we confront ever more difficult choices. The first approach we shall examine is an outgrowth of the ethics of freedom.

11.1 INDIVIDUALISM AND FREEDOM

In the West, the modern conception of humankind is characterized, more than anything else, by individualism. The implications of this individualism give rise to our understanding of freedom. One

> *No man can put a chain about the ankle of his fellow man without at last finding the other end fastened about his own neck.*
>
> *Frederick Douglass[a].*

[a]Frederick Douglas - speech, Civil Rights Mass Meeting, Washington, D.C., 1883.

Figure 11.2. Douglass.

effort to fully explore the notion of freedom can be found in existentialist theory[1]. An existentialist conception of individuality may give rise to the following set of questions relevant for our search for approaches to confronting serious questions in the 21st century:

- What is human freedom?

- What can the absolute freedom of absolute individuals mean?

- What is human flourishing or human happiness?

- What general ethic or way of life emerges when we take our individuality seriously?

- What ought we to do?

- What ethics or code of action can emerge from a position that takes our individuality seriously?

Sartre[2] explores an appropriate ethics code using existentialist theory. According to Sartre, we each individually choose human nature for all humans. Hence, we must choose courses of action that we would wish all humans to take. In choosing for ourselves, we choose for all of humanity. Thus, I must choose in the same way we would want others to choose - another instance of the use of the Golden Rule. We speak of acting authentically when we ignore the external differences among ourselves and other people as these differences are merely outward manifestations of who we are —

[1]Existentialism is a philosophy that emphasizes the uniqueness and isolation of the individual experience in a hostile or indifferent universe, regards human existence as unexplainable, and stresses freedom of choice and responsibility for the consequences of one's acts.

[2]Jean Paul Sartre (1905-1980) was a French novelist, playwright, existentialist philosopher, and literary critic. Sartre was awarded the Nobel Prize for literature in 1964, but he declined the honor in protest of the values of bourgeois society.

not the essence of who we are. Sartre also argues that in order to be free, we must desire the freedom of all humanity. It is self-defeating to attempt to use other humans as objects to satisfy our desires, or to protect our freedom at the cost of enslaving others. The person who uses other people as objects to satisfy his desires makes himself or herself an object. To see others as slaves of our desire is to make ourselves a slave of desire. Thirdly, our decisions are not arbitrary as we speak of a coherence of our actions. Our actions must unify the many different influences on our lives into the one life that is to be ours. Our actions, though free, are constrained by our situation in a community with all its relationships and obligations.

In summary, freedom as constructed from an existentialist perspective, must take on the responsibility of choosing for all of humankind, desire and work for the freedom of all humanity, and create ourselves within the context of the relationships and obligations we have to others.

11.2 ETHICS OF FREEDOM: SUBSTANTIAL FREEDOM

Sen[3] is seen as a ground-breaker among late 20th century economists for his insistence on discussing issues seen as marginal by most. He mounted one of the few major challenges to the economic model (capitalism) that has placed self-interest as the prime motivating factor of human activity. Sen (Amartya, 2005) describes what he refers to as the basic idea of positive or substantial freedom, distinguishes it from negative freedom on the one side and happiness on the other, and relates it to the notion of "capability," which is distinct from a raw capacity and an actual exercise of a capability. He further states that "the perspective of freedom" is concerned with "enhancing the lives we lead and the freedoms we enjoy." The ethic of freedom calls for, "expanding the freedoms we have reason to value," so that our lives will be "richer and more unfettered" and we will be able to become "fuller social persons, exercising our own volitions and interacting with–and influencing–the world in which we live."

Garret (2008) offers the following clarification of what Sen means by 'substantial freedom.'

According to Garret, "Substantial freedoms are valuable things that can be divided up and delivered to human beings (or groups of people in a region) in varying amounts. In that respect they are like money and freedom from coercion, things which can be preconditions for substantial freedom but are not very good indicators of it. In the case of money, a person can have very little, a middle amount, or a lot. In the case of freedom from coercion, the same can be said: one can be a slave, constantly subject to the whims of an overseer, or one can have maximum available freedom from coercion by one's fellow humans in their private or governmental capacities. What society does, and to some extent what individuals do, can determine how much substantial freedom we have."

Continuing, "substantial freedom is distinguished from other things that we 'often have reason to value:' money, negative freedom or freedom from coercion, and happiness, on the other. Monetary income alone cannot be used as a reliable indicator of substantial freedom. An increase in income

---

[3] Amartya Kumar Sen is an Indian economist, philosopher, and a winner of the Nobel Prize for Economics in 1998, "for his contributions to welfare economics" for his work on famine, human development theory, welfare economics, the underlying mechanisms of poverty, and political liberalism.

might be converted into an increase in substantial freedom, but the conversion is not automatic or equally easy for everybody. A sick person is normally less able than a healthy one to convert a given increase in income into a wider range of real opportunities, i.e., into greater substantial freedom. The same might be said of a person who lives in a dangerous neighborhood that makes him/her fearful to go outside as compared to a person who lives in a safer neighborhood" (Garret, 2008).

11.3   ETHICS OF FREEDOM: CAPABILITIES

Nussbaum (1999) has continued to develop the notion of substantial freedom. According to Nussbaum, "At the heart of this tradition is a twofold intuition about human beings: namely, that all, just by being human, are of equal dignity and worth, no matter where they are situated in society, and that the primary source of this worth is a power of moral choice within them, a power that consists in the ability to plan a life in accordance with one's own evaluation of ends." To these two ideas is linked one more, that "the moral equality of persons gives them a fair claim to certain types of treatment at the hands of society and politics. This treatment must do two things: respect and promote the liberty of choice, and …respect and promote the equal worth of persons as choosers."

A necessary component of Nussbaum's approach is the list of basic capabilities. She answers the question, "What activities characteristically performed by human beings are so central that they seem definitive of a life that is truly human?" Nussbaum's list includes the following:

- The ability to live life to its natural end

- Maintaining health and integrity of the body

- The ability to move freely about and be free from the threat of violence

- Being able to use the senses; being able to imagine, to think, and to reason

- Being able to have attachments to things and persons outside ourselves; being able to love those who love and care for us while not having one's emotional developing blighted by fear or anxiety.

- Being able to form a conception of the good and to engage in critical reflection about the planning of one's own life.

- Being able to live for and in relation to others, to recognize and show concern for other human beings, to engage in various forms of social interaction; being able to imagine the situation of another and to have compassion for that situation; having the capability for both justice and friendship.

- Being able to be treated as a dignified being whose worth is equal to that of others.

- Being able to live with concern for and in relation to animals, plants, and the world of nature.

- Being able to laugh, to play, to enjoy recreational activities.

- Possessing control over one's environment.

Challenge Box: Suppose you were challenged to put an engineering ethic based on freedom in your own words and have it be less than fifty words. What would you write? What similarities would there be between this approach versus the other more traditional approaches we have already taken? What differences?

Let's go back to our trusted trolley problem. What would an ethic based on freedom point to for both scenarios? Explain. Are the recommendations different than what you concluded earlier?

CHAPTER 12

Engineering and Chaos

Figure 12.1. Chaos.

In the twentieth century, science is undergoing a major reshuffling. The work of Einstein and others has shown that Newtonian science adequately describes a limited number of idealized problems at best. A new science, the science of chaos, is beginning to supplant the classical mechanics of Newton. It is a science of disorder, probabilities and non-linearities. Most interestingly, it appears to be a more accurate description of nature, for nature's essence is chaos. The winds of the atmosphere, the streams of the oceans, the sliding of the surface platelets all are chaotic. Chaos also describes the deposition of river silt along the Mississippi Delta, the flow of water through the giant oak trees,

I accept chaos. I am not sure whether it accepts me. I know some people are terrified of the bomb. But then some people are terrified to be seen carrying a modern screen magazine. Experience teaches us that silence terrifies people the most.

Bob Dylan.

Figure 12.2. Dylan.

You need chaos in your soul to give birth to a dancing star.

Nietzsche.

Figure 12.3. Nietzsche.

and the flow of blood in the arteries and veins of human beings. The nonlinear interaction of one human-being with another is also chaotic.

12.1   DEFINITION

What is chaos? In fact, many pundits or 'talking-heads' that appear regularly in the media claim our world is becoming ever more chaotic so its probably useful for us to understand what is meant by the term. The notion of chaos is linked to change which we can categorize into a few fundamental categories: growth and recession, stagnation, cyclic behavior and unpredictable, erratic fluctuations. Nearly all of these phenomena can be described with well developed *linear* mathematical tools. Here *linear* means that the result of an action is always proportional to its cause: if we double our effort, the outcome will also double. Most of nature is non-linear in the same sense as most of zoology is non-elephant zoology (Ulam, 0000). Scientists who study chaos and its broader are of specialty, non-liner mechanics, often liken the situation that most of traditional science is focusing on linear systems can be compared to the story of the person who looks for the lost car keys under a street lamp because it is too dark to see anything at the place where the keys were lost. With recent advances in computational speed, memory and computing algorithms, scientists now have the ability to make significant progress in the field of non-linear systems and begin to understand phenomena that we had no hope of understanding in the past.

This phenomenon, *chaos*, exists outside the framework of linear theory. The modern notion of *chaos* describes irregular and highly complex structures in time and in space that follow deterministic laws and equations which is in contrast to the totally random chaos of classical thermodynamics.

Mayer-Kress offers the following visualization (Figure 12.4) to illustrate the difference among determinism, chaos and randomness (Mayer-Kress, 1995).

"Consider a fluid in a pot on a stove in which the level of stress is given by the rate at which the fluid is heated. At room temperature, the water is in equilibrium with its surroundings is totally placid. The appearance of placidity indicates a total lack of spatial or temporal coherence. As heat is added to the water, the system starts to self-organize and form regular spatial patterns (rolls, hexagons) which create coherent behavior of the subsystems ('order parameters slave subsystems'). The order parameters themselves do not evolve in time. Under increasing stress the order parameters themselves begin to oscillate in an organized manner: we have coherent and ordered dynamics of the subsystems. Further increase of the external stress leads to bifurcations to more complicated temporal behavior, but the system as such is still acting coherently. This continues until the system shows temporal deterministic chaos. The dynamics is now predictable only for a finite time. This predictability time depends on the degree of chaos present in the system. It will decrease as the system becomes more chaotic. The spatial coherence of the system will be destroyed and independent subsystems will emerge which will interact and create temporary coherent structures."

This flow field is termed a Benard flow field in which convection cells that appear spontaneously in a liquid layer when heat is applied from below. In this configuration, we have a thin layer of water or some other liquid filling the gap between two parallel plates. The bottom plate is then heated. A simple sketch of the resulting flow field is shown in Figure 12.5.

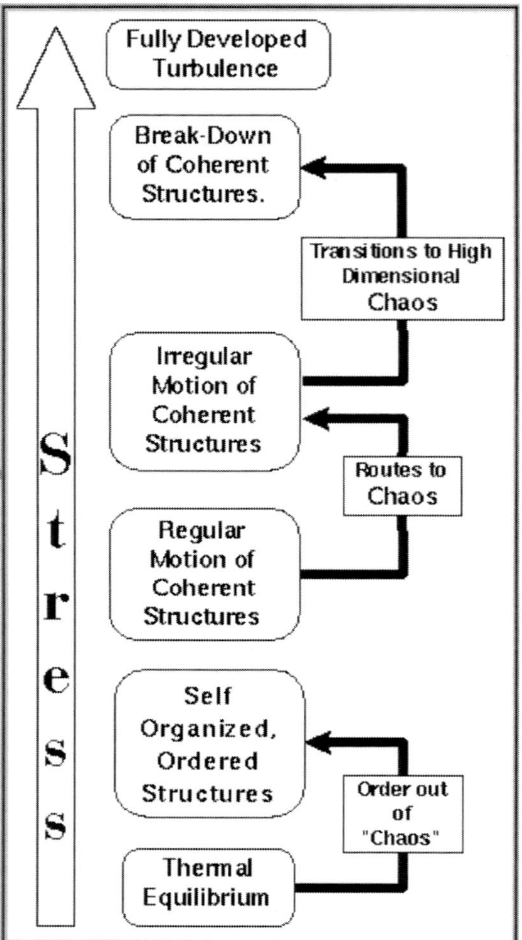

Figure 12.4. Onset of turbulence: chaos to order & order to chaos (Mayer-Kress, 1995).

12.2   NATURE AND CHAOS

Our efforts to understand the shift in perspective offered by chaos may be best illustrated by first considering the implications chaos has for our understanding of the natural world. After exploring these implications, we will ask the following questions: How does chaos affect the way I may approach decision-making in the 21st century? What new insights does it offer?

Nature still was thought to be "naturally" in an equilibrium condition. An example of the Newtonian view of the stability and orderliness of the natural world is to be found in the writings of George Perkins Marsh (Marsh, 2008).

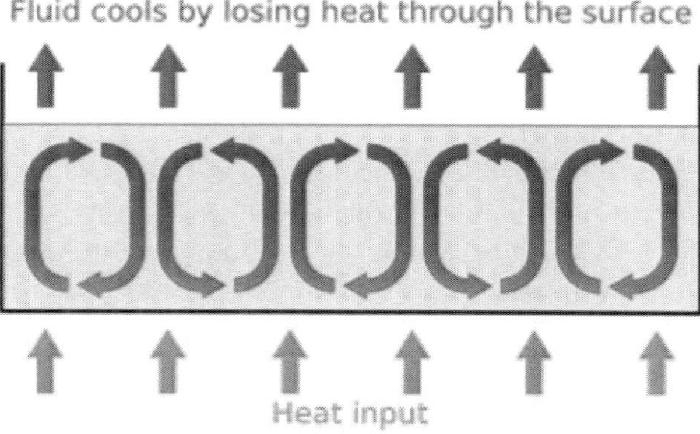

Figure 12.5. Schematic of a Benard Cell.

Nature, left undisturbed so fashions her territory as to give it almost unchanging permanence, outline, and proportion, except when shattered by geologic convulsions; and in these comparatively rare cases of derangement, she sets herself at once to repair the superficial damage, and to restore, as nearly as practicable, the former aspect of her dominion.

Newtonian mechanics views the universe as a gigantic mechanism that functions like clockwork, and represents the ultimate in reliability and mechanical perfection. Newtonian mechanics casts the laws of nature in the form of mathematical equations (Stewart, 1998). The key point about these equations is that each of their solutions is unique. For example, a bicycle has perhaps some four or five moving parts. If the motion of each of the parts one instant of time is known, then Newtonian mechanics allows the determination of the motion of the bicycle at some later point farther down the road. Cosmically, the knowledge of the position and the velocities of every point of matter in the Solar System at one instant in time would allow the unique determination of all subsequent motions of those particles at some later time. This statement both assumes for mathematical simplicity that there are no external influences on this motion, and leads to the interpretation that the positions and the velocities of every particle in the entire universe taken at some fixed instant of time completely determines its future evolution. The entire universe follows a unique, predetermined path according to the French mathematician Laplace [from Stewart (1998)]:

"An intellect which at any given moment knew all the forces that animate Nature and the mutual position of the beings that comprise it, if this intellect were vast enough to submit its data to analysis, could condense into a single formula the movement of the greatest bodies of the universe and that of the lightest atom: for such an intellect nothing would be uncertain; and the future just like the past would be present before its eyes."

This statement by Laplace describes the paradigm of classical determinism. The paradigm worked well in the solutions of many technical problems and provided the mathematical foundation for the West's "Industrial Revolution." There was great optimism, for it seemed only a matter of time until all the problems of classical mechanics would be solved. Yet, there remained one problem in Newtonian mechanics: how to describe mathematically the turbulent motion of a fluid.

Turbulent flow describes the motion of the natural world. Nearly all motion that occurs in nature is turbulent. Turbulence is nonlinear, dynamic, random, multi-dimensional, and is characterized by indeterminacy and intermittency. Turbulent motion is the motion of the clouds, the earth's winds and seas, and the flow of living fluids through the major arteries of living creatures (plants and animals alike). Turbulence also describes not only the interaction of one member of a species with another, but also the interaction of one species with another. Prior to the latter half of the twentieth century, scientists believed that the differential equations, independently developed by Navier and Stokes, would effectively predict the laminar flow of a fluid and would also prove adequate for the challenge of turbulent flow. They anticipated that the advent of faster and larger computers or ever more esoteric and *ad hoc* models would provide the missing link in the search for the deterministic solution.

Such hopes appear today to be naive at best or arrogant at worst, for a deterministic solution for the turbulent flow of a fluid is no nearer at hand than it was at the turn of the 20th century. Rather than the old paradigm of classical determinism, a paradigm based on chaos more clearly describes turbulence. While Newtonian mechanics enables the determination of exact solutions to a narrow set of idealized problems, it has not been adequate in scientists' attempts to model and predict turbulent fluid flow. Rather than being an uncommon occurrence, the turbulent motion of a fluid is the motion of life and reflects the essence of the natural world. The new science of chaos holds potential for a clearer understanding of turbulence and, hence, a clearer understanding of the natural world.

In a fluid, we have turbulent cascades where vortices are created that will decay into smaller and smaller vortices. Analog situations in societies can be currently studied in the former USSR and Eastern Europe. James Marti speculates: "Chaos might be the new world order" (Mayer-Kress, 1995). At the limit of extremely high stress, we are back to an irregular chaos where each of the subsystems can be described as random and incoherent components without stable, coherent structures. It has some similarities to the anarchy with which we started close to thermal equilibrium. Thus, the notion of *chaos* covers the range from completely coherent, slightly unpredictable, strongly confined, small scale motion to highly unpredictable, spatially incoherent motion of individual subsystem.

12.3   NEWTONIAN AND CHAOS-BASED ETHIC

In *Sand County Almanac* (Leopold, 1968), Leopold wrote:

> "All ethics so far evolved rest upon a single premise: that the individual is a member of a community of interdependent parts. Ecology simply enlarges the boundaries to include soils, waters, plants, and animals, or collectively the land. In short a land ethic changes

the role of Homo *Sapiens* from conqueror of the land community to plain member and citizen of it. It implies respect for his fellow members and also respect for the community as such."

Leopold went on to formulate "The Land Ethic:"

A thing is right when it tends to preserve the integrity, stability, and beauty of the biotic community. It is wrong when it tends otherwise.

Leopold's view of the natural world is apparent when he asserts that the *stability of Nature is disrupted by the interference of Man.* He believed that Nature left to its own design will "naturally" opt for the stable or equilibrium position and that it is Man who disrupts this stability, challenges its integrity, and causes a natural world with diminished beauty. This view is a clear precursor of the thought of Thomas Berry, who compared the beauty of the Hudson River before its occupation by the Western world to its present condition and found its grandeur diminished.

The failure of the mechanical model of the natural world can be demonstrated through various examples[1]. Sometimes more than simply being folly, the Newtonian view of nature led to ecological disasters.

If we shift our model of the natural world from its historic deterministic base to one which embraces chaos, a new ethic can be suggested:

A thing is right when it tends to allow the natural world and all the entities thereof, to thrive in richness and diversity, and to experience change. It is wrong when it tends otherwise.

Note that there are three elements to this new ethic.

- Richness here refers to the richness of experience of the various entities that make up the natural world (Johnson, 1991).

- Diversity refers to wide variety in plant and animal species. An action which would result in the enhancement of the variety of species that would exist in a given ecosystem would be in accord with the new environmental ethic.

- Change as the restraint of change is a violation of the processes that model the natural world.

[1] Probably no other controversy has done more to divide the ranks of conservationists around the world or more to cripple ecological research in East Africa than that involving the elephants of Tsavo. Because of pressures outside the huge 8,300 square-mile Tsavo National Park, thousands of elephants have been crowding into the sanctuary, swelling the population to somewhere between 20,000 and 30,000, and making it the largest remaining concentration of elephants in the world. Because elephants eat so much and push down so many trees in the course of their activities, large concentrations can devastate an area. Tsavo, once a lush land of dense bush and trees is today a sad landscape of bare ground and struggling grasses, strewn with the skeletons of downed trees. Because the park receives very little rain, some scientists have said the destruction could eventually turn Tsavo into a desert, providing so little food that most of the elephants would eventually starve to death. The long-term result could be a "population crash" that would wipe out elephants in the very place set aside to protect them. Three years ago a drought reduced the already damaged vegetation so severely that between 5,000 and 6,000 elephants died of malnutrition. Since then, still more elephants have entered the park, seeking refuge from hunters and expanding agriculture. The population may again be as high as before the drought. Further droughts and massive die-offs are almost certain in coming months and years. When a scientific team recommended shooting 3,000 elephants for research purposes and indicated it might be necessary to shoot many more to bring the population back into balance with its environment, many conservationists recoiled in horror. Killing the animals they were trying to protect hardly fit the traditional conservation ethic.

Challenge Box: Compare and contrast a vision of a mechanical universe versus one that is based on chaos. Which one makes the most sense according to your values/beliefs? What similarities would there be between this chaos based approach versus the ethic of freedom and other more traditional approaches we have already taken? What differences? Let's go back to our trusted trolley problem. What would an ethic based on chaos point to for both scenarios? Explain. Are the recommendations different than what you concluded earlier?

C H A P T E R 13

Engineering and a Morally Deep World

Figure 13.1. Berry.

> *Geese appear high over us, pass, and the sky closes. Abandon, as in love or sleep, holds them to their way, clear in the ancient faith: what we need is here. And we pray, not for new earth or heaven, but to be quiet in heart, and in eye, clear. What we need is here.*
>
> *Wendell Berry.*

A new code of ethics is offered for engineering adapted from an environmental model of nature as a self-organizing system. A self-organizing system is characterized by synthesis rather than analysis and suggests a new code of ethical responsibility based upon community rather than individuality.

13.1 SELF-ORGANIZING SYSTEMS

Modern science at the start of the 21st century does not model the natural world using either the great chain of being or the mechanical clock paradigms or as living being (Gaia hypothesis). Today, the natural world is most often described using the model of a self-organizing system and nature rather than being thought of as immutable is seen as constantly in change. Self-organization refers to a process in which the internal organization of a system, normally an open system automatically without being guided or managed by an outside source. Self-organizing systems typically (though not always) display emergent properties. Emergence is the process of complex pattern formation from simpler rules. This can be dynamic (occurring over time), such as the evolution of the human brain over thousands of successive generations; or emergence can happen over disparate length scales, such as the interactions between a macroscopic number of neurons producing a human brain capable of thought (even though the constituent neurons are not themselves conscious). For a phenomenon to be termed emergent, it should generally be unpredictable from a lower level description.

The world abounds with systems and organisms that maintain a high internal energy and organization in seeming defiance of the laws of physics. According to Decker (1995), "As a bar of iron cools, ferromagnetic particles magnetically align themselves with their neighbors until the entire bar is highly organized. Water particles suspended in air form clouds. An ant grows from a single-celled zygote into a complex multicellular organism, and then participates in a structured hive society. What is so fascinating is that the organization seems to emerge spontaneously from disordered conditions, and it does not appear to be driven solely by known physical laws. Somehow, the order arises from the multitude of interactions among the simple parts. The laws that may govern this self-organizing behavior are not well understood, if they exist at all. It is clear, though, that the process is nonlinear, using positive and negative feedback loops among components at the lowest level of the system, and between them and the structures that form at higher levels."

Decker adds:

> "The study of landscape ecology provides an example of how an SOS perspective differs from standard approaches. Ecologists are interested in how spatial and temporal patterns such as patches, boundaries, cycles, and succession arise in complex, heterogeneous communities. Early models of pattern formation use a 'top-down' approach, meaning the parameters describe the higher hierarchical levels of the system. For instance, individual trees are not described explicitly, but patches of trees are. Or predators are modeled as a homogenous population that uniformly impacts a homogeneous prey population. In this way, the population dynamics are defined at the higher level of the population, rather than being the results of activity at the lower level of the individual."

Finally,

> "The problem with this top-down approach is that it violates two basic features of biological (and many physiochemical) phenomena: individuality and locality. By modeling a rodent population as a mass of rodents with some growth and behavior parameters, we obviate any differences that might exist between individual rodents. Some are big, some are small, some reproduce more, and some get eaten more. These small differences can lead to larger differences - such as changes in the population gene frequencies, individual body size, or population densities - that might have cascading effects at still higher levels. The tenet of locality means that every event or interaction has some location and some range of effect. This is a simple illustration of the ecological principle that pattern affects process. To say that a system is self-organized is to say it is not entirely directed by top-down rules, although there might be global constraints on the system. Instead, the local actions and interactions of individuals generate ordered structures at higher levels with recognizable dynamics. Since the origins of order in SOS are the subtle differences

among components and the interactions among them, system dynamics cannot usually be understood by decomposing the system into its constituent parts. Thus, the study of SOS is synthetic rather than analytic."

If the self-organized system is used to model the natural world rather than the great chain of being or the mechanical clock, our sense of responsibilities to the natural world seem to change significantly. We are forced to look "synthetically" rather than "locally," that is at the very least or moral sphere of concern must broaden. Secondly, nature is no longer in perfect order nor is it a collection of parts, i.e., gears, levers, weight) which can be replaced or modified according to our desires. The mechanical clock in many ways has been replaced by a seemingly chaotic clock which defies predictability, single-valueness and repeatability. If we are to make sense of our place in this natural world, we need a very different sense of ethics. One attempt at providing such an ethical framework has been offered by Johnson (1993) in his development of a morally deep world.

13.2 A MORALLY DEEP WORLD

According to Leopold, acting ethically is a matter of concern both for us and for others with whom we are in some sort of community. The notion of a community deserves some discussion. We, perhaps, are most comfortable with community referring to a body of people having common rights, privileges, or interests, or living in the same place under the same laws and regulations; as, a community of Franciscan monks. In biology or ecology, community refers to an interacting group of various species in a common location. For example, a forest of trees and undergrowth plants, inhabited by animals and rooted in soil containing bacteria and fungi, constitutes an integral community (Johnson, 1993). Extending the notion of community in this way is consistent with the pattern evidenced in human society over the centuries. We have progressively enlarged the boundaries of our understanding of community and recognized the membership of slaves, foreigners, etc., those for whom membership was not extended at earlier times in history. Leopold's land ethic then "simply enlarges the boundaries of the community to include soils, waters, plants, and animals, or collectively: the land."

Johnson discusses how non-sentient land can count morally and focuses upon the concept of a living being (Johnson, 1993). For Johnson, a living being is best thought of not as a thing of some sort but as a living system, an ongoing life-process. A life-process has a character significantly different from those of other processes such as thermodynamics processes, for example. Our character, as living beings, is the fundamental determinant of our interests. Johnson adds further that:

> "The interests of a being lie in whatever contributes to its coherent effective functioning as an on-going life-process. That which tends to the contrary is against its interests....moral consideration must be given to the interests of all living beings, in proportion to the interest. Some living systems other than individual organisms are living entities with morally considerable interests. ...All interests must be taken into account."

The concept of a morally deep world was developed within the framework of environmental ethics. Perhaps, it may be useful to explore the morally deep world argument as it applies to a specific and presently quite contentious issue in wildlife management today, the reintroduction of the Mexican wolf into regions of the Southwestern United States. For the purposes of illustration, let us focus on the land near the White Sands Missile Range near Las Cruces, New Mexico. Johnson would challenge us to first identify all the members of the community. For this example, a listing would include the following:

- Wolves

- Prey animals including domestic sheep and cattle as well as deer, rabbits, coyotes, and others

- Desert lands

- Ranchers and sheep farmers

- Hunters

- U.S. Fish and Wildlife Service and other state and local government agencies

- U.S. Department of Defense

- Residents of White Sands and nearby towns and settlements

- Residents of New Mexico and the entire United States

- Native American residents

Often in such cases, two very different perspectives dominate the deliberations. On one side of the debate is atomism, a view that moral assessment applies only to individuals. The individual would be individual wolves, prey, ranchers, etc. On the other side is holism, a view that collectives or whole are subject to moral appraisal. In a morally deep world, the view is shortsighted morally if one adopts either a holistic or atomistic. No one (holistic or atomistic) interest has priority over the other. There is an inevitable tension between atomistic and holistic ethics. Sometime the interests of the biotic community will outweigh the interests of the individual, while at other times it is the interests of the individual which are paramount. Let us next identify the extent of the community or living being in this case. Recall that a living being is characterized as having an ongoing life process with interests in whatever contributes to its coherent effective functioning. Clearly, wolves, their prey, the desert lands, ranchers, sheep farmers, hunters and people who live in or near White Sands have considerable interests. Other identified elements could be argued to have less interest in the coherent effective functioning of the community. That is not to suggest that, for example, the residents in New York would have no interest in the restoration, but their impact on the coherent effective functioning of the ongoing process would be less.

An interesting example of the tension between atomism and holism can be identified in the following scenario. Suppose wolves are restored to the White Sands Missile range desert and suppose that, as has been the case in Yellowstone National Park, wolves adapt well and quickly grow in numbers. In Yellowstone, some wolves are routinely killed as part of wolf or game management practices. From a holistic perspective this may be morally acceptable, but it would be difficult to justify the killing from an atomistic perspective. A morally deep world point of view would argue that both interests need to be considered carefully, including the interests of the entire park community and those of the "surplus" wolf.

One criticism often offered of a morally deep world perspective is that it prevents any action that will affect a community. On the contrary, though a morally deep perspective does assert actions that violate vital interests of the community or erosion of its self-identity should be avoided, it requires active participation in the protection of the essential functions and the maintenance of the viability of life processes. Rather than calling for inaction, a morally deep world perspective suggests contemplation followed by direct and specific responses.

13.3 ENGINEERING IN A MORALLY DEEP WORLD

Given a shift to a morally deep world paradigm, a new engineering code of ethics is outlined. The majority of existing codes are structured, in similar if not identical ways, with fundamental principles supported by fundamental canons. That same structure will be incorporated into the present work. For a morally deep world, the first fundamental canon and rule of practice is specified as:

Engineers, in the fulfillment of their professional duties, shall hold paramount the safety, health and welfare of the identified integral community.

The fundamental difference between an ethical code based on a morally deep world versus the present codes is the replacement of the "public" by the "identified integral community."

How would decisions be made in engineering adhering to a morally deep world perspective? The following flow chart is offered with the decision process consisting of the following four steps:

- Via Positiva. The problem is identified, fully accepted and broken down into its various components using the vast array of creative and critical thinking techniques which engineers possess. What is to be solved? For whom is it to be solved?

- Via Negativa. Reflection on the possible implications and consequences for any proposed solution are explored. What are the ethical considerations involved? The societal implications? The global consequences? The effects on the natural environment?

- Via Creativa. The third step refers to the act of creation. The solution is chosen from a host of possibilities, implemented and then evaluated as to its effectiveness in meeting the desired goals and fulfilling the specified criteria.

- Via Transfomativa. The fourth and final step asks the following questions of the engineer: Has the suffering in the world been reduced? Have the social injustices that pervade our global village been even slightly ameliorated? Has the notion of a community of interests been expanded? Is the world a kinder, gentler place borrowing from the Greek poet Aeschylus? (Smith, 1970).

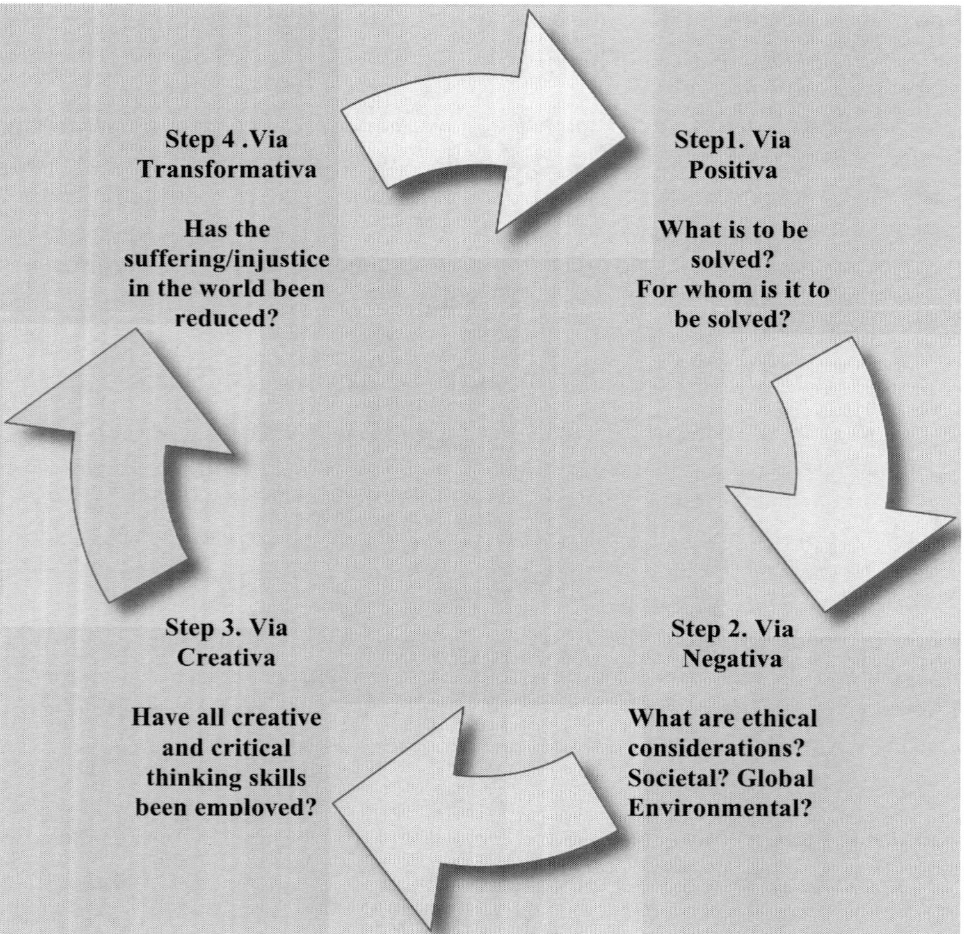

Figure 13.2. Engineering decision algorithm based upon a Morally Deep World.

13.4 APPLYING THE MORALLY DEEP WORLD VIEW

Consider the following scenario. Harvesting machines are replacing migrant workers as grape pickers in Northern California (Tesconi, 2000). Today, such machines cost approximately $126,000. It takes five workers about eight hours to pick 10 tons of grapes. To harvest the 100 tons would require 90 workers if they wanted to be done before sun-up. The costs are relatively straightforward to calculate.

It costs $120 a ton to harvest by hand, and $30 a ton by machine. A mid-size wine producer expects to harvest about 4,000 tons of grapes total. For 4,000 tons, the cost to pick the grapes by hand is approximately $480,000 while the costs using a harvester are $120,000 representing an increase in margin above cost equal to $340,000. Even with the cost of the harvester subtracted from this margin, the net increase in profits is equal to $204,000.

Traditional engineering ethics discussions might stop here. But what if we go a bit farther and identify the workers as being part of the community? What else would we be forced to consider? If it takes 90 workers to harvest 100 tons then it would take the equivalent of 3600 workers to harvest 4000 tons. Certainly there are many more people involved than simply the workers themselves with the total involved growing far beyond the estimate of 3600. What will become of them? Should we as the engineers who designed, built, tested, evaluated and delivered the harvester care? If we use the code of ethics described by countless engineering societies today, the answer would be no. There is no consideration given to the workers whose livelihoods have been eliminated. Codes of engineering ethics based on the notion of a morally deep world would suggest a very different result as it forces us as engineers to consider the entire living system, in this case, including the men, women and children who toil as grape pickers and whose quality of life is intimately linked to the harvest.

Figure 13.3. An example of a mechanical fruit picker.

For this case, the following questions concerning the significance of the grape-harvesting device may be asked:

- For whom should the device be designed? The owners? The land? The workers? Someone else?

- What will become of the displaced workers?

- What will become of the displaced workers families?

- Are there long-term effects on the land?

- Are there societal implications for the workers' communities?

- Are there societal implications for the greater community outside the workers and their families?

The identified members of the community may include:

- Owners

- Workers

- Workers' families

- Land and water ecosystem

- Farm worker society

- Local, regional and national societies

Suppose we use the proposed engineering design algorithm. In Via Positiva, an identification of the problem is made. Grapes have to be picked at a particular time and in a particularly rapid fashion. The harvesting of grapes by hand has in fact been done satisfactorily for thousands or years on farms of all sizes and now to a limited extent by machines. In Via Negativa, the possible implications or consequences of the change to a harvesting machine are considered. The effects upon the environment for the given case seem minimally affected by the mechanization of the process as stated here. The consequences for the workers, their families and their ways of life are however profound and disturbing. A large number of workers will be displaced and their jobs completely eliminated. In Via Creativa, engineers have, in fact, designed and delivered machines that will meet the criteria set forth by the grape farmers and have thus met their professional responsibilities as has been commonly understood. But a consideration of the net effects on the suffering in the world, Via Transformativa, might yield a different result. Engineers who design such devices without concern for the impact on the numbers of farm workers who are being replaced solely in order to increase the profits of the landowners may have acted unethically.

In the context of the farm workers discussed here, a logical question to ask would be the following: Is it possible to arrive at an end result which was a creative design that met everyone's needs and established justice for workers and their families and furthered the interests of the vineyard? While we cannot offer a device, we can suggest that those whom the new machine would make irrelevant might be included in discussions concerning the consequences of implementing the new design. The landowner may ultimately decide to mechanize this task but at the very least the criteria for making ethical choices would also include a careful consideration of the impact of the engineering design on the lives of the workers. It would seem such an inclusion is as important as any other.

Challenge Box: Summarize in your own words what is meant by a morally deep world. Suppose you were challenged to write down a morally deep world ethic for engineering in less than 100 words, what would you write?

What similarities would there be between this morally deep world approach versus other more traditional approaches we have already taken? What differences?

Let's go back to our trusted trolley problem. What would an ethic based on a morally deep world point to for both scenarios? Explain. Are the recommendations different than what you concluded earlier?

C H A P T E R   14

Engineering and Globalism

It is yet another *Civilized Power, with its banner of the Prince of Peace in one hand and its loot-basket and its butcher-knife in the other.*

Mark Twain (1992).

Figure 14.1. Mark Twain.

There is little doubt that the world is becoming more and more globalized. The current form of globalization which is based upon neo-liberalism, free trade and open markets has sparked much debate throughout the world. Some critics would argue that the interests of powerful nations and corporations are shaping the terms of world trade. Proponents would argue that globalization has led to massive increases in the world's wealth while critics would suggest that while a few people are becoming increasingly wealthy, a greater percentage of the world's population is become poorer.

14.1 GLOBALISM AND GLOBALIZATION

Globalism, at its core, seeks to describe and explain a world which is characterized by networks of connections that span multi-continental distances. It attempts to understand all the inter-connections of the modern world — and to highlight patterns that underlie and explain them. In contrast, globalization refers to the increase or decline in the degree of globalism. It focuses on the forces, the dynamism or speed of these changes. Globalism is the underlying basic network, while globalization refers to the dynamic shrinking of distance on a large scale.

Globalization in its literal sense is the process of making, transformation of some things or phenomena into global ones. It can be described as a process by which the people of the world are unified into a single society and function together. This process is a combination of economic, technological, socio-cultural and political forces. Globalization is often used to refer to economic globalization, that is, integration of national economies into the international economy through trade, foreign direct investment, capital flows, migration, and the spread of technology.

Friedman (2007) "examines the impact of the 'flattening' of the globe," and argues that "globalized trade, outsourcing, supply chaining , and political forces have changed the world permanently, for both better and worse. He also argues that the pace of globalization is quickening and will continue to have a growing impact on business organization and practice. In the book, Friedman recounts a journey to India when he realized globalization has changed core economic concepts. He suggests the world is "flat" in the sense that globalization has leveled the competitive playing fields between industrial and emerging market countries. In his opinion, this flattening is a product of a convergence of personal computer with fiber-optic micro cable with the rise of work flow software. He termed this period as Globalization 3.0, differentiating this period from the previous Globalization 1.0 (which countries and governments were the main protagonists) and the Globalization 2.0 (which multinational companies led the way in driving global integration)."

Chomsky[1] argues that the word globalization is also used, in a doctrinal sense, to describe the neo-liberal form of economic globalization. Chomsky (1995) surprisingly agrees with Adam Smith[2] that at the outset that:

> "free movement of people is a core component of free trade. As for free movement of capital, that's a totally different matter. Unlike persons of flesh and blood, capital has no rights, at least by Enlightenment/classical liberal standards. As soon as we bring up the matter of free movement of capital, we have to face the fact that while people are in principle at least equal in rights, in a just society, talk of capital conceals the reality: we are speaking of owners of capital, who are vastly unequal in power, naturally. In the real world, free movement of capital entails radical restriction of democracy, for obvious reasons that have long been well understood. Speaking of capital and labour as if they were on a par is so hopelessly misleading that sensible discussion is impossible in these terms."

He develops the term "just globalization" in which he describes globalization as he thinks it should be when issues of democratic versus authoritarian control of production, distribution, interchange, information, etc., are all carefully considered and weighed in the conversations.

Daly suggests that sometimes the terms internationalization and globalization are used interchangeably but there is a slight formal difference (Daly, 2000). The term "internationalization" refers to the importance of international trade, relations, treaties etc. International means between or among nations. "Globalization" means erasure of national boundaries for economic purposes; international trade becomes inter-regional trade.

---

[1]Noam Chomsky is an American linguist, philosopher, political activist, author and lecturer at the Massachusetts Institute of Technology.

[2]Adam Smith (1723-1790) was a Scottish moral philosopher and a pioneering political economist. Smith is known for his explanation of how rational self-interest and competition, operating in a social framework which ultimately depends on adherence to moral obligations, can lead to economic well-being and prosperity. He is widely acknowledged as the "father of economics."

There are four distinct dimensions of globalism: economic, military, environmental and social:

- Economic globalism involves long-distance flows of goods, services and capital and the information and perceptions that accompany market exchange.

- Environmental globalism refers to the long-distance transport of materials in the atmosphere or oceans or of biological substances such as pathogens or genetic materials that affect human health and well-being.

- Military globalism refers to long-distance networks in which force, and the threat or promise of force, are deployed.

- The fourth dimension is social and cultural globalism. It involves movements of ideas, information, images and of people, who carry ideas and information with them.

Discussion on globalism primarily focuses on economic globalism. This phenomenon seems particularly important in the practice of engineering and we shall explore the implications of this form and suggest an ethical framework for making decisions in light of its existence.

With respect to economic globalism, the number of people living in absolute poverty has increased from a billion five years ago to 1.2 billion today according to a collaborative report prepared by the World Bank, the International Monetary Fund, the Organization for Economic Cooperation and Development and the United Nations. For more than 30 of the poorest national economies, real per capita incomes have been falling for the past 35 years. Asia is the only region in which poverty rates decreased during the past five years. Economic progress in Latin America was made ineffective by the increase in inequality among the various classes of society. People in the industrial countries now are 74 times richer than those in the poorest. The wealth of the three richest men in the world is greater than the combined GNP of all of the least developed countries - 600 million people. This impoverishment has occurred at a time when globalization was supposed to have launched the poor into sustained economic growth.

14.2   EMERGENCE OF A GLOBAL ETHIC

On September 4, 1993, for the first time in the history of religion, delegates to the Parliament of the World's Religions in Chicago adopted a "Declaration toward a Global Ethic"[3]. On September 1, 1997, again for the first time, the InterAction Council of former heads of state or government

[3]Declaration towards a Global Ethic, Chicago, 1993. The world is in agony. The agony is so pervasive and urgent that we are compelled to name its manifestations so that the depth of this pain may be made clear. Peace eludes us – the planet is being destroyed – neighbors live in fear – women and men are estranged from each other – children die! This is abhorrent. We condemn the abuses of Earth's ecosystems. We condemn the poverty that stifles life's potential; the hunger that weakens the human body, the economic disparities that threaten so many families with ruin. We condemn the social disarray of the nations; the disregard for justice which pushes citizens to the margin; the anarchy overtaking our communities; and the insane death of children from violence. In particular, we condemn aggression and hatred in the name of religion. But this agony need not be. It need not be because the basis for an ethic already exists. This ethic offers the possibility of a better individual and global order, and leads individuals away from despair and societies away from chaos. We are women and men who have embraced the precepts and practices of the world's religions: We affirm that a common set of core values is found in the teachings of the religions, and that

called for a global ethic and submitted to the United Nations a proposed "Universal Declaration of Human Responsibilities," designed to underpin, reinforce and supplement human rights from an ethical angle. In addition, the third Parliament of the World's Religions, held in Cape Town in December 1999 issued "A Call to Our Guiding Institutions," based on the Chicago Declaration.

What is, in fact, meant by a *global ethic?* By Global Ethic is meant the necessary minimum of common values, standards and basic attitudes or alternatively a minimal basic consensus relating to binding values, irrevocable standards and moral attitudes, which can be affirmed by all religions despite their undeniable dogmatic or theological differences and also be supported by non-believers. In many ways, it is a call for a change of consciousness. In its widest sense it includes all our sensations, thoughts, feelings, and volitions—in fact, the sum total of our mental life.

Challenge Box: Summarize in your own words what is meant by global ethic. Suppose you were challenged to write down a global ethic for engineering in less than 100 words, what would you write?

What similarities would there be between this global ethic approach versus other more traditional approaches we have already taken? What differences?

Let's go back to our trusted trolley problem. What would a global ethic point of view point to for both scenarios? Explain. Are the recommendations different than what you concluded earlier?

these form the basis of a global ethic. We affirm that this truth is already known, but yet to be lived in heart and action. We affirm that there is an irrevocable, unconditional norm for all areas of life, for families and communities, for races, nations, and religions. There already exist ancient guidelines for human behavior which are found in the teachings of the religions of the world and which are the condition for a sustainable world order. We declare: We are interdependent. Each of us depends on the well-being of the whole, and so we have respect for the community of living beings, for people, animals, and plants, and for the preservation of Earth, the air, water and soil. We take individual responsibility for all we do. All our decisions, actions, and failures to act have consequences. We must treat others as we wish others to treat us. We make a commitment to respect life and dignity, individuality and diversity, so that every person is treated humanely, without exception. We must have patience and acceptance. We must be able to forgive, learning from the past but never allowing ourselves to be enslaved by memories of hate. Opening our hearts to one another, we must sink our narrow differences for the cause of the world community, practicing a culture of solidarity and relatedness. We consider humankind our family. We must strive to be kind and generous. We must not live for ourselves alone, but should also serve others, never forgetting the children, the aged, the poor, the suffering, the disabled, the refugees, and the lonely. No person should ever be considered or treated as a second-class citizen, or be exploited in any way whatsoever. There should be equal partnership between men and women. We must not commit any kind of sexual immorality. We must put behind us all forms of domination or abuse. We commit ourselves to a culture of non-violence, respect, justice, and peace. We shall not oppress, injure, torture, or kill other human beings, forsaking violence as a means of settling differences. We must strive for a just social and economic order, in which everyone has an equal chance to reach full potential as a human being. We must speak and act truthfully and with compassion, dealing fairly with all, and avoiding prejudice and hatred. We must not steal. We must move beyond the dominance of greed for power, prestige, money, and consumption to make a just and peaceful world. Earth cannot be changed for the better unless the consciousness of individuals is changed first. We pledge to increase our awareness by disciplining our minds, by meditation, by prayer, or by positive thinking. Without risk and a readiness to sacrifice there can be no fundamental change in our situation. Therefore, we commit ourselves to this global ethic, to understanding one another, and to socially beneficial, peace-fostering, and nature-friendly ways of life. We invite all people, whether religious or not, to do the same.

C H A P T E R   15

Engineering and Love

Figure 15.1. Love.

Love and compassion are necessities, not luxuries.
Without them, humanity cannot survive.

If the love within your mind is lost and you see other beings as enemies, then no matter how much knowledge or education or material comfort you have, only suffering and confusion will ensue.

Dalai Lama (1998).

> *Morals were too essential to the happiness of man, to be risked on the uncertain combinations of the head. [Nature] laid their foundation, therefore, in sentiment, not in science.*
>
> *Thomas Jefferson.*

Figure 15.2. Jefferson.

Engineering based on love? We have encountered the more traditional applied ethics, an ethic based on freedom, another on chaos and still others based on a morally deep world view or globalism. Each offers us different ways to respond when confronted with ethical dilemmas. Our last approach to confronting difficult decisions will be an engineering ethic based on love. Key elements of an engineering ethic based upon love would include the capacity for true, rigorous critical thought, the development of a culture in which individual dissent is honored and revered and in which each of us considers our self a citizen of the Earth. Lastly, an engineering ethic would enable each of us to develop our own individual narrative of moral imagination, that is, to develop the ability to be in another's shoes, to cultivate our inner eye of seeing and knowing and to overcome the blindness that we have all become far too accustomed. First, we shall explore the definitions of love used in this exploration.

15.1 DEFINITIONS OF LOVE

Countless sages, scholars, poets, philosophers, theologians and others have tried to define love throughout the ages. We shall use the following description of three aspects of love which may impact the manner in which we reach decisions.

- *Agape*: love that promotes overall well-being when confronted by that which generates ill-feeling (i.e., returning good for ill)

- *Eros*: love that promotes overall well-being by affirming the valuable or beautiful

- *Philia*: love that promotes overall well-being when cooperating with others.

15.2   AN ETHIC OF LOVE

Putting the three ideas together, we have the basic framework for reaching ethical decisions. Our work then must promote the overall well-being of all, including our perceptions of friends, and foes. It challenges us to reflect on the words found in nearly all universal wisdom traditions. One example is given below:

> *But I say to you that hear, love your enemies, do good to those who hate you, bless those who curse you, pray for those who abuse you. To those who strike you on the cheek offer the other also, and from those who take away your cloak, do not withhold your coat as well. Give to everyone who begs from you, and of those who take away your goods, do not ask them again. And as you wish that others would to do you, do so to them (Luke 6, ESV Bible, 2008).*

It calls for moving beyond that false dualism, false as who is to say what actually is good and what is ill? Can we ever really know with certainty? Making decisions, using a sports metaphor, ought not to be equivalent to cheering for one side over another whether it is American football or that variety of football they play in every other corner of the world.

Secondly, how can we affirm the beautiful and the valuable? Perhaps the first step is to examine that which we view as beautiful and that which we view as valuable. The subjective experience of "beauty" often involves the interpretation of some entity as being in balance and harmony with nature which may lead to feelings of attraction and emotional well-being. In its most profound sense, beauty may engender a significant or important experience of positive reflection about the meaning of one's own existence. An "object of beauty" is anything that reveals or resonates with personal meaning. Inner beauty is a concept used to describe the positive aspects of something that is not physically observable. Qualities including kindness, sensitivity, tenderness or compassion; creativity and intelligence have been said to be desirable since antiquity. In turn, let us consider what each of these qualities is asking from us.

- Kindness: the act or the state of charitable behavior to other people.

  > *"To give pleasure to a single heart by a single kind act is better than a thousand head-bowings in prayer."*
  > *Saadi*

- Sensitivity: the quality or condition of being sensitive, that is, the capacity to respond to stimulation.

  > *"Either I'm too sensitive, or else I'm gettin' soft,"*
  > *Bob Dylan*

- Tenderness: the quality or state of being considerate or protective.

  > *"Courage is by no means incompatible with tenderness. On the contrary, gentleness and tenderness have been found to characterize the men, no less than the women, who have*

done the most courageous deeds."
Samuel Smiles

- Compassion: the human feeling of pity over another's sorrows, along with the desire to help others in their situations.

   "A human being is a part of the whole called by us universe, a part limited in time and space. He experiences himself, his thoughts and feeling as something separated from the rest, a kind of optical delusion of his consciousness. This delusion is a kind of prison for us, restricting us to our personal desires and to affection for a few persons nearest to us. Our task must be to free ourselves from this prison by widening our circle of compassion to embrace all living creatures and the whole of nature in its beauty."
   Albert Einstein

- Creativity: the ability to see something in a new way, to see and solve problems no one else may know exists, and to engage in mental and physical experiences that are new, unique, or different.

   "The world is but a canvas to the imagination."
   Henry David Thoreau

- Intelligence: a property of mind that encompasses many related mental abilities, such as the capacities to reason, plan, solve problems, think abstractly, comprehend ideas and language, and learn.

   It's not that I'm so smart; it's just that I stay with problems longer.
   Albert Einstein

Affirming what we view as valuable at the outset requires that we clarify what we mean when we say we value some one or some idea. In essence, what we are doing is clarifying what we value, and why we value it. A sevenfold process describing the guidelines of the values clarification approach was formulated by Simon et al. (1995).

- choosing from alternatives;

- choosing freely;

- prizing one's choice;

- affirming one's choice;

- acting upon one's choice; and

- acting repeatedly, over time.

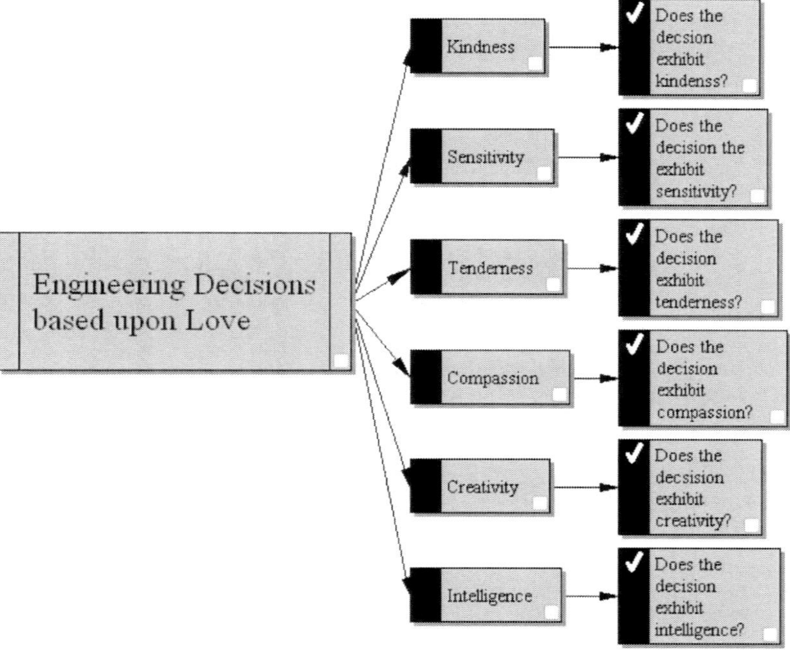

Engineering Decisions based upon Love

Figure 15.3. Seven elements of engineering decisions based upon love.

Additional theorists providing support for the values clarification approach include Asch (1952) and G. Murphy (1958).

The last category of love, *philia*, challenges us to see the world in a different way, in the words of Thomas Berry (2006) as a *collection of subjects*. Berry's most famous quotation is:

The Universe and thus the Earth is a communion of subjects, not a collection of objects.

By communion, Berry was referring to intimacy or a feeling of emotional closeness, a connection, especially one in which something is communicated or shared. The shift from object to subject[1] is also profound. An object is something visible or tangible; something that can be seen or touched, a focus of somebody's attention or emotion; or a goal or purpose. By subject, the reference is to the essential nature or substance of something as distinguished from its attributes. In other

---

[1] In Physicist Conception of Nature, Werner Heisenberg's underlines the fundamental chance in the status of subject/object relationship, brought about by the quantum theory (and the Copenhagen interpretation of it). Newtonian physics has a clear-cut distinction between object and subject. When an XIX century physicist was approaching the study of the nature, he was hoping to unveil the law of it; the subject of his study was nature "itself." After quantum physics this is no longer possible - there is no "nature itself." The process of observation for ever changes the observed. The observer and the observed are interacting. Heisenberg writes: "We can no longer speak of the behavior of the particle independently of the process of observation." The laws we formulate are not about the nature itself, but about our knowledge of it.

words, borrowing from Buddhism,[2] the essential nature, the Buddha nature, is taught to be a truly real, but internally hidden, eternal potency or immortal element within the purest depths of the mind, present in all sentient beings. When we practice our profession of engineering, it is important that we view humanity and the ecosystem as part of an undivisable whole. Berry takes this one step farther. According to Berry, our new community is a very special one, that is, it is one in which the various elements are bound together as subjects having interests rather than one in which some have interests while others are simply resources to be utilized.

In our view, viewing the Universe including both the natural environment and the poor as a communion of subjects rather than a collection of objects has important, even revolutionary significance for the engineering profession. First, it eliminates from the outset that we ever again can remain aloof from the consequences of our projects. Landmines continue to explode long after the end of hostilities in combat zones across the globe. Polar bears are rapidly disappearing from the Arctic regions in part due to the technologies we continue to produce. The poor in New Orleans suffered beyond our understanding in part due to decisions we as engineers and engineering organizations made and continue to routinely make. If we can begin to see the connection we have with the health of the Arctic ecosystem and thus with the well-being of the polar bears, recognizing all that we share; they like the rest of nature have much greater importance when we are formulating our criteria whereby we make decisions. Even more importantly, if we can begin to view the poor, rather they live in the 9th Ward of New Orleans or the Pine Ridge Reservation in South Dakota, as connected to us and as possessing an entire spectrum of potentialities and possibilities then too our criteria for decision making as engineers is broadened importantly. Those potentialities and possibilities are as important to the ongoing dynamic process of creation in the Universe as those that reside within us as each of us plays an integral role in the communion of subjects.

In engineering, we often speak of development. Far too often, it seems that the model used in engineering is linked solely to profit making. The ultimate goal is economic growth with no interest in peace, social or environmental justice or wealth distribution. Such a model ignores inequalities, has contempt for the arts and literature, promotes group think, needs docile practitioners and de-emphasizes critical thinking. I would like to offer a different paradigm for engineering, one which has as its priority the development of not only the human spirit but also the rest of the natural world. Using such a paradigm, each and every being matters, groups are disaggregated into individuals and equal respect exists for each individual. Ultimately, the goal of such a profession would be to enable each of us to transcend our own particular situations and imagine a global society which is based upon equality and on love. Key elements of our profession based upon love would include the capacity for true, rigorous critical thought, the development of a culture in which individual dissent is honored and revered, and in which each of us considers our self a citizen of the Earth. Lastly, an engineering based upon love would enable each of us to develop our own individual narrative of

---

[2] Buddhism is a dharma, non-theistic religion, a philosophy, and a system of psychology. Buddhism is also known in Sanskrit or Pail, the main ancient languages of Buddhists, as Buddha Dharma which means the teachings of "the Awakened One."

moral imagination, that is, to develop the ability to be in another's shoes, to cultivate our inner eye of seeing and knowing and to overcome the blindness that we have all become far too accustomed.

Challenge Box: Summarize in your own words what is meant by an engineering ethic based on love. Suppose you were challenged to write down such an ethic for engineering in less than 100 words, what would you write?

What similarities would there be between ethical approach versus other more traditional approaches we have already taken? What differences?

Let's go back to our trusted trolley problem. What would this ethical point of view point to for both scenarios? Explain. Are the recommendations different than what you concluded earlier?

C H A P T E R   16

Case Study Application

We shall conclude with several different case studies, each representative of the kinds of real-world decision each of you, as engineers, may confront in your upcoming professional careers. The present text has laid out a range of options whereby you can carefully consider the cases in light of applied ethical theories ranging from those traditionally used in engineering to new ideas based on developments in moral reasoning to include issues such as freedom, chaos, a morally deep world view, and globalism, and ultimately, one based on love. After the case is described, some possible suggestive questions will be provided as a means for you to begin your own, personal reflection on the various cases and your responses as a future engineering professional.

16.1   THE NEXT GENERATION OF LAND MINES

Part I.

Ms. Jane Enaj is a project manager at a multinational corporation which as just been awarded a contract to develop and produce the next generation land mine. She is also a member of the Design Review Committee. The committee's responsibilities include reviewing and approving design changes, procedural changes and submitting performance reports to various U.S. Department of Defense agencies with recommendations.

Today, Jane finds herself in a difficult situation. DRC is meeting to finalize recommendation concerning the new land mine. It offers significant improvements for the U.S. military as it will self-detonate in a set period of time thereby reducing the risk of having unexploded ordinance remain on the battlefield long after the actual conflict has ended. This new design should reduce the number of injuries and deaths to innocent civilians by a large amount and is set for deployment in the Middle East. Yet innocent citizens including children and the elderly will still be maimed and/or die. In addition, unsuspecting pets, domestic animals and wildlife will also likely suffer.

The unit has returned after a two month trail run in various environments including U.S desert southwest, and island possessions in the Pacific. After extensive engineering analysis by engineers, it has been concluded that the new design falls slightly above the minimum requirement set by the design specifications. The test and analysis results were both promising and disappointing as approximately 50% of the mines self-destructed. This still would represent a significant improvement over the use of the land mines in the present U.S. arsenal.

Seven members of DRC are present, enough for a quorum. Ms. Enaj is the least senior member present. From the outset of the meeting, committee chair Mr. Senior has made it clear that it is important to act quickly since any delay will cost the company, and a lot of money. "A redesign," he says, "might take several months. If we don't approve this, we may be facing a multi-million dollar loss in revenues. We have met the design specifications. What do you think?" Mr. Smith and Mr. Jones immediately concur. Mr. Senior then says, "Well, if no one sees any problems here, let's go with it." There is a moment of silence. Suppose you were Ms. Enaj. What would you recommend?

Part II.

Right after the design meeting a breakthrough was made in the design of the actual explosive material used in the mine. For the same amount of explosives material used, the magnitude of the explosion is doubled both in intensity and radius of impact. The DRC calls a second meeting to consider what impact the new explosives material may have on their final recommendations. Again playing the role of Ms. Enaj, what would you recommend and/or do?

Part III.

After the land mine has been put into use by American forces, the Chinese and Pakistani governments submit orders for significantly large number of the new devices. China and Pakistan are in turn is known to sell similar weapon systems to various regimes throughout the world. Again playing the role of Ms. Enaj, what would you recommend and/or do?

16.2 ADVANCES IN AUTOMOTIVE TECHNOLOGIES IN THE DEVELOPING WORLD

Part I.

Mr. Sam Mas is a project engineer for the electronics branch of a multinational automotive company. Mr. Mas' design team has recently designed, produced and tested a new control module for possible use in a mass produced economy car. The new module has the potential for reducing the manufacturing cost as well as increasing the energy efficiency of the car. The new design is ideally suited as part of the company's strategy to increase sales in under-developed nations throughout the world. The new car has the potential to revolutionize transportation in regions of the globe where travel is still quite expensive. The company's board of directors is meeting in order to decide on the future for this new module. The plans are to produce the first version of the module in northern California and then subsequently to transfer production to Bangladesh to reduce labor and manufacturing costs. Mr. Mas is asked to provide his recommendations to the board as they consider implementation of the new system. Acting in the role of Mr. Mas, what would you recommendations to the board be?

Part II.

The new module has been put into production with the production facilities transferred to Bangladesh as planned. The sale of automobiles has increased dramatically in less affluent countries. The production facility in the U.S. has been abandoned with the resultant dislocation of the workers involved. The neighborhood in which the production facility has been located has seen a rapid

Next Generation of Landmines

Utilitarianism	Rights of Persons
□ What are all the available options for Ms. Enaj *concerning* the design review? □ What are the costs of delaying the introduction of the landmine? □ Who will benefit the most from its introduction? □ What costs are associated with increased explosive capabilities? □ Who will benefit the most from its increased explosive capabilities? □ What decisions maximize the good? □ What are the costs associated with the selling of the weapons system to China and Pakistan?	□ What are the effects of the improved land mine on the combatants and on civilians? □ Would we be willing to serve as soldiers under such an arrangement? As civilians? □ Would we be able to accept the consequences of the new landmine on ourselves as soldiers? □ Would we be able to accept the consequences of the new landmine on ourselves as innocent civilians?
Virtue	Freedom
□ Are the rights of all involved in the landmine production and use given their just due? □ Does the new design and its implementation demonstrate self-control and/or discipline? □ Does the design and use of the new landmine design require moral strength and/or courage? □ Does its design and use signify a pursuit of good according to reason? □ Or is it simply a manifestation of our fears?	□ In designing and using the new landmine design are we cognizant of the responsibility we have to all of humankind? □ Are we working for the freedom of all of humanity? □ Are we aware of the context of relationships that exist? □ Are we cognizant of the obligations we have to ourselves as well as to others? □ Are we aware of our individuality and take it in to account?

Figure 16.1. Next generation of landmines: suggested reflective questions.

transformation away from the original native culture. The company's board of director is meeting again to consider even greater increases in control module production. Mr. Mas is invited to provide engineering's perspective. Acting in the role of Mr. Mas, what would you recommendations to the board be?

Next Generation of Landmines

Chaos	Morally Deep World
□ Are we taking into account in our landmine design the integrity of the biotic community? □ Are we considering the stability of the same community as well as the stability of the cultures involved? □ Are we cognizant of the beauty of the local environment, different cultures including our own? □ Are we enriching the potential experiences of the various elements within the communities? □ Are we respecting and promoting diversity of not only species but also thought and custom? □ Are we arbitrarily seeking to restrain change? □ Are we comfortable with uncertainty?	□ For whom are we designing the landmine? □ Whose interests does it serve? □ Have we identified the integral community for the design? □ Who is included in the integral community? □ What other elements are involved in the integral community? □ Have we called on our vast array of creativity as well as analytical skills? □ have we considered the possible negative consequences involved in the mine design? □ Has the pain and suffering in the world increased, decreased, remained the same or unknown as our design is put into theaters around the world?
Globalism	Love
□ Do we consider the global ethic when we design the new landmine? □ Are we cognizant of the interdependence of all life and the planet? □ Are we promoting a more equitable order with the landmine use? □ Is there a chance that more will benefit as a result of its use? □ Or does it serve the self-serving interests of a few? □ Have we tried in many different ways to fully understand the basis for the conflict itself? □ Are we treating those with whom we come into contact with our devices in the same way we would wish to be treated?	□ Does the new design exhibit any elements of kindness? □ To those who must use it? □ To those who will ultimately be maimed or killed? □ Is it sensitive to the needs of not only the combatants but also the innocent civilians and the ecosystem? □ Does the device foster a sense of protectiveness? For whom? □ As designers, have we demonstrated compassion for whom will be affected by our device? □ Have we been as creative as we can in its design, execution, implementation? □ Have we ben willing to think outside the box, to use our mind to its fullest capabilities?

Figure 16.2. Next generation of landmines: suggested reflective questions.

16.3   LEVEES, COASTLANDS, AND SAFETY

Part I.

Dr. Zeeman, a career governmental employee, is the chief engineering officer for the government agency overseeing the design and reconstruction of the levees in and around the New Orleans

Advances in Automotive Technology

Utilitarianism	Rights of Persons
□ What are all the available options for Mr. Mas concerning the design review? □ What are the costs of delaying the introduction of the module? □ Who will benefit the most from its introduction? □ What costs are associated with increased sales of autos? □ Who will benefit the most from its increased sales and use? □ What decisions maximize the good? □ What are the costs associated with moving the plant to Bangladesh?	□ What are the effects of increased use of autos? □ Would we be willing to accept more autos in our towns? Our country?? □ Have we considered the impact of changes to transportation to the native culture(s)? □ Would we be able to accept the consequences of the new landmine on ourselves as innocent civilians? □ Have we considered the effects of job relocation on those originally employed in the U.S.?
Virtue	Freedom
□ Are the rights of all involved in the control production and use given their just due? □ Does the new design and its implementation demonstrate self-control and/or discipline? □ Does the design and use of the new control module require moral strength and/or courage? □ Does its design and use signify a pursuit of good according to reason? □ Or is it simply a manifestation of our selfishness or greed?	□ In designing and using the new module design are we cognizant of the responsibility we have to all of humankind? □ Are we working for the freedom of all of humanity? □ Are we aware of the context of relationships that exist? Among the workers in the U.S.? Among the workers in Bangladesh? □ Are we cognizant of the obligations we have to ourselves as well as to others? □ Are we aware of our individuality and take it in to account?

Figure 16.3. Automotive technology: suggested reflective questions.

area. His agency has performed an in-depth and careful analysis of the failure of the levees as a result of Hurricane Katrina. The final report provides a design for three different scenarios: a Category 3, a Category 4 and a Category 5 strength hurricane striking the metropolitan area of New Orleans. The new design focuses on strengthening the levee system as it presently exists. A meeting of Dr. Zeeman as well as appropriate governmental officials is called with the purpose to finalize plans for redesign. A budget crisis has reduced the amount of funds available for the project with the result

Advances in Automotive Technology

Chaos	Morally Deep World
▫ Are we considering the integrity of the biotic community with new auto production and use? ▫ Are we considering the stability of the same community as well as the stability of the cultures involved with the increased use? ▫ Are we cognizant of the beauty of the local environment, different cultures including our own? ▫ Are we enriching the potential experiences of the various elements within the communities? ▫ Are we respecting and promoting diversity of not only species but also thought and custom? ▫ Are cognizant of the various societies, their customs, and their histories?	▫ For whom are we designing the control module? ▫ Whose interests does it serve? ▫ Have we identified the integral community for the design? ▫ Who is included in the integral community? ▫ What other elements are involved in the integral community? ▫ Have we called on our vast array of creativity as well as analytical skills? ▫ Have we considered the possible negative consequences involved in the module design? ▫ Has the pain and suffering in the world increased, decreased, remained the same or unknown as our design is put into use around the world?
Globalism	Love
▫ Do we consider the global ethic when we increase use of cars? ▫ Are we cognizant of the interdependence of all life and the planet? ▫ Are we promoting a more equitable order with more auto use? ▫ Is there a chance that more will benefit as a result of its use? ▫ Or does it serve the self-serving interests of a few? ▫ Have we tried in many different ways to fully understand the basis for the conflict itself? ▫ Are we treating those whose lives are changing due to the design, production and relocation of production in the same way we would wish to be treated?	▫ Does the new design exhibit any elements of kindness? ▫ To those who must use it? ▫ To those who will ultimately be affected by it? ▫ Is it sensitive to the needs of not only the producers, the users as well as the bystanders? ▫ Does the device foster a sense of protectiveness? For whom? ▫ As designers, have we demonstrated compassion for whom will be affected by our device? ▫ Have we been as creative as we can in its design, execution, implementation? ▫ Have we been willing to think outside the box, to use our mind to its fullest capabilities?

Figure 16.4. Advances in automotive technology: suggested reflective questions.

being. Dr. Zeeman is asked to sign off the plan to insure that the levees can withstand a Category 3 hurricane. Suppose you were the take the place of Dr. Zeeman at the meeting, would you do?

Part II.

A study from a nearby land grant university points to the accelerating erosion of the Louisiana coastland as a result of the levees system. This study comes to your attention immediately after the planning meeting on the Category 3 design implementation. The loss of coastal wetlands increases the risk of damage from subsequent hurricanes as well as the destruction of the local ecosystem with loss of plants, wildlife and income for those residents who make their income from the area. Again acting in the role of Dr. Zeeman, what would you do?

Part II.

The levee reconstruction goes ahead as planned with no attention paid to the issue of population evacuation. The plans in place have not change since Hurricane Katrina. Your agency is not responsible for developing an evacuation plan. A local newspaper consults you on a story involving post-Katrina readiness and safety. How would you respond to the reporter's questions?

16.4 NATIONAL PARKS, DANGEROUS DRIVERS, AND THE REMOVAL OF TREES

Part I.

Isabella Arbole is the Chair of the Road Safety Commission for the Rocky Mountain National Park (RSCRMNP). Her agency has primary responsibility for maintaining the safety of all roads within the confines of the park. Visitors to the park have increased by 200% in the past 10 years. This has resulted in increased traffic flow on both primary and secondary roads in the area. Alpine Drive, still a two lane road, has more than tripled its traffic flow during this period.

For each of the past 10 years at least one person per year has suffered a fatal automobile accident by crashing into trees closely aligned along a 5 mile stretch of Alpine Drive. Many other accidents have also occurred, causing serious injuries, wrecked cars, and damaged trees. Some of the trees are quite close to the pavement. Two law suits have been filed against the park for not maintaining sufficient road safety. Both cases were decided in the government's favor as not enough evidence was submitted to prove the government's responsibility for the crashes as in each case the drivers were traveling way above the posted speed limit and the road conditions were dangerous due to freak storms.

Other members of RSCRMNP have been pressing Ms. Arbole to propose a solution to the traffic problem on Alpine Drive. They are concerned about safety, as well as law suits that may some day go against RSCRMNP. Suppose you are acting as Ms. Arbole's engineering expert, what would you recommend?

Part II.

A significant increase in the elk population occurs in the park. The increase while not threatening the balance of the ecosystem does pose a significant problem for safety on the roadways. The numbers of deaths as a result of automobile-elk crashes has increased from less than one per year to on average three per year. The other members of RSCRMNP have been pressing Ms. Arbole to propose a solution. They are again concerned about safety, as well as law suits that may some day go against RSCRMNP. Suppose once again you are acting as Ms. Arbole's engineering expert, what would you recommend?

Levees, Coast lands and Safety

Utilitarianism	Rights of Persons
□ What are all the available options for the levee redesign? □ What are the costs associated with the design? □ Who will benefit the most from its completion? □ What costs are associated with increased coastal erosion? □ Who will benefit from the status quo? □ What decisions maximize the good? □ What are the evacuation considerations and costs associated with the new levee system?	□ What are the effects of the new design on past, present and future residents? □ Would we be willing to move/live in New Orleans under such an arrangement? □ Would we be able to accept the consequences of the new design? □ Would we be able to accept the consequences of the new design on ourselves as innocent civilians? □ Would we be willing to live in New Orleans realizing that we may not be able to evacuate?
Virtue	Freedom
□ Are the rights of all involved in the levee redesign given their just due? □ Does the new design and its implementation demonstrate self-control and/or discipline? □ Does the design require evidence of moral strength and/or courage? □ Does the design and use signify a pursuit of good according to reason? □ Or is it simply a manifestation of our fears? Or biases? Or prejudices?	□ In designing and using the new levee design are we cognizant of the responsibility we have to all of humankind? □ Are we working for the freedom of all of humanity? □ Are we aware of the context of relationships that exist? □ Are we cognizant of the obligations we have to ourselves as well as to others? □ Are we aware of our individuality and take it into account?

Figure 16.5. Levees, coastlands and safety: suggested reflective questions.

Levees, Coast lands and Safety

Chaos	Morally Deep World
□ Are we considering the integrity of the biotic community in New Orleans? □ Are we considering the stability of the same community as well as the stability of the cultures involved with the increased use? □ Are we cognizant of the beauty of New Orleans? □ Are we enriching the potential experiences of the various elements within the communities? □ Are we respecting and promoting diversity of not only species but also thought and custom? □ Are cognizant of the various societies, their customs, and their histories?	□ For whom are we redesigning the levee system? □ Whose interests does it serve? □ Have we identified the integral community for the design? □ Who is included in the integral community? □ What other elements are involved in the integral community? □ Have we called on our vast array of creativity as well as analytical skills? □ Have we considered the possible negative consequences involved in proposed levee redesign? □ Has the pain and suffering in the world increased, decreased, remained the same or unknown for the various elements of the integral community?
Globalism	Love
□ Do we consider the global ethic when we redo the levee system? □ Are we cognizant of the interdependence of all life and the planet? □ Are we promoting a more equitable order with more auto use? □ Is there a chance that more will benefit as a result of its use? □ Or does it serve the self-serving interests of a few? □ Have we tried in many different ways to fully understand the implication of the design and its omissions? □ Are we treating those whose lives are changing due to the design in the same way we would wish to be treated?	□ Does the new design exhibit any elements of kindness? □ To those who must use it? □ To those who will ultimately be affected by it? □ Is it sensitive to the needs of not only the producers, the users as well as the bystanders? □ Does levee system provide a sense of protectiveness? For whom? □ As designers, have we demonstrated compassion for whom will be affected by our design? □ Have we been as creative as we can in its design, execution, implementation? □ Have we been willing to think outside the box, to use our mind to its fullest capabilities?

Figure 16.6. Levees, coastlands and safety: suggested reflective questions.

National Parks, Dangerous Drivers and Trees

Utilitarianism	Rights of Persons
□ What are all the available options for Ms. Arbole concerning the design ? □ What are the costs of new roads and tree removal in the park? Of delaying the new design? □ Who will benefit the most from its introduction? □ What costs are associated with increased use? □ Who will benefit the most from its increased use? □ What decisions maximize the good? □ What are the costs associated with the increased use, tree removal, elk harvesting?	□ What are the effects of the improved roads on the visitors? On local residents? □ Would we be willing to serve as live under such an arrangement? A □ Would we be able to accept the consequences of the increased visitation, use and speed of travel? □ Would we be able to accept the consequences of the new system on ourselves, our communities, and our local ecosystems?
Virtue	Freedom
□ Are the rights of all involved in the road design their just due? □ Does the new design and its implementation demonstrate self-control and/or discipline? □ Does the design require moral strength and/or courage? □ Does its design and use signify a pursuit of good according to reason? □ Or is it simply a manifestation of our fears? □ Or is it simply a manifestation of our greed? □ Or is it simply a manifestation of our narcissism?	□ In designing and using the transportation design are we cognizant of the responsibility we have to all of humankind? □ Are we working for the freedom of all of humanity? □ Are we aware of the context of relationships that exist? □ Are we cognizant of the obligations we have to ourselves as well as to others? □ Are we aware of our individuality and take it in to account?

Figure 16.7. National parks, dangerous drivers, and the removal of trees: suggested reflective questions.

National Parks, Dangerous Drivers and Trees

Chaos	Morally Deep World
□ Are we considering the integrity of the biotic community in Rocky Mountain National Park? □ Are we considering the stability of park as well as the stability of the cultures involved with the increased use? □ Are we cognizant of the beauty of park? □ Are we enriching the potential experiences of the various elements within the park? □ Are we respecting and promoting diversity of not only species but also thought and custom? □ Are cognizant of the various ecosystems, societies, their customs, and their histories?	□ For whom are we redesigning the road system in the park? □ Whose interests does it serve? □ Have we identified the integral community for the design? □ Who is included in the integral community? □ What other elements are involved in the integral community? □ Have we called on our vast array of creativity as well as analytical skills? □ Have we considered the possible negative consequences involved in proposed road redesign? □ Has the pain and suffering in the world increased, decreased, remained the same or unknown for the various elements of the integral community?
Globalism	**Love**
□ Do we consider the global ethic when we redo the road system in the park? □ Are we cognizant of the interdependence of all life in the park and the planet? □ Are we promoting a more equitable order ? □ Is there a chance that more will benefit as a result of more and safer roads? □ Or does it serve the self-serving interests of a few? □ Have we tried in many different ways to fully understand the implication of the design and its omissions? □ Are we treating those whose lives are changing due to the design in the same way we would wish to be treated?	□ Does the new design exhibit any elements of kindness? □ To those who must use it? □ To those who will ultimately be affected by it? The trees? The elk? □ Is it sensitive to the needs of all? □ Does road system provide a sense of protectiveness? For whom? □ As designers, have we demonstrated compassion for whom will be affected by our design? □ Have we been as creative as we can in its design, execution, implementation? □ Have we been willing to think outside the box, to use our mind to its fullest capabilities?

Figure 16.8. National parks, dangerous drivers and the removal of trees: suggested reflective questions.

Final Thoughts

I am the voice crying in the wilderness...the voice of Christ in the desert of this island...[saying that] you are all in mortal sin...on account of the cruelty and tyranny with which you use these innocent people. Are these not men? Have they not rational souls? Must not you love them as you love yourselves?

Antonio de Montesinos (1987).

The aim of the present text is to provide each of you with the skills to be able to confront the challenges that awaits us as professional engineers in the 21st century. The world is becoming ever more complicated, ever more interconnected. It is our belief that it only serves to help us if we are aware of a wide range of possible approaches we may take when confronted with the need to make decisions today and in the future. We are challenging you and the discipline of engineering to continue to embody an ethical approach to issues of safety, and professionalism and the like but also to take a much broader view of our macro-ethical responsibilities. That is why we have provided you with new tools of applied ethical analysis to include notions of freedom, of chaos, of a morally deep world view, of globalism and ultimately of love. This is an important shift in the engineering profession and in our view one that is desperately needed.

In many ways, the shift that we are calling for is remarkably similar to an impassioned call for a radical shift in consciousness in Western thought that occurred almost 500 years earlier in Spain. It is a debate that brought to the forefront many of the same issues we are dealing with now in the 21st century. The confrontation was known as the Valladolid debate and it concerned the treatment of natives of the New World. Held in the city of Valladolid in Spain, the debate featured the two main attitudes taken in Spain towards the conquests of the New World. On one side, Dominican monk Bartolome de Las Casas argued that the natives were free men in the natural order and deserved the same treatment as others, according to Catholic theology. Las Casas was opposed by fellow Dominican monk, Sepulveda, who insisted the Indians were natural slaves, and therefore reducing them to slavery or serfdom was in accordance with Catholic theology and natural law.

The debate focused on questions such as: What are just wars? When is violence justified? What are the responsibilities of the developed world towards the under-developed world? What does it mean to be a human being? What rights do all human being possess by virtue of being human? What rights if any does the planet possess?

Sadly, no positive outcome came out of the debate; no realistic solution could have resulted, for the debate was carried out in too theoretical a framework. Both sides, determined to prove or disprove the legality of war as a means of conversion, adamantly stuck to their respective writings, and thus failed to reach a realistic and concrete compromise. Not surprisingly, the debate failed to materialize into palpable benefits for the Indians.

Though concrete steps to alleviate the suffering of the Indians did not result as a result of the debate, that event signaled the beginning of a long journey towards greater awareness of the plight of the Indians but also of other groups such as those ensnared in the African slave trade. Consciousness was raised. It was a beginning of a conversation, a conversation that continues to this day. Our hope is that in some small way our work will result in the beginnings of a conversation and a similar awakening of consciousness in the profession of engineering today as we move ever farther into the 21st century.

Part III

Engineering: Windows on Society

CHAPTER 18

Introduction

This volume is intended to be read in conjunction with the first two volumes of the series, 'Engineers and Society' and 'Making Decisions in the 21st Century.' We present a series of every day case studies, which exemplify the issues these volumes raise and we look at these cases through different lenses. It is possible to read these cases without first having studied the previous volumes although it is recommended that they be consulted at some point. In the first volume, it was explained that in these texts we take what is known as a critical perspective. In case you have not read the first volume, we will repeat some of the key features of this way of presenting material. If you have read Volume 1, skip the rest of the introduction and move straight to the cases.

Taking a critical perspective does not mean that we will criticise – but which we will critique through different lenses. 'Critical Theory' is an amalgam of philosophical and social scientific techniques providing a way to systematically, critically and yet constructively analyse systems of thought and practice. It includes a variety of possible approaches and perspectives by which to analyse 'not only cultural artefacts but also their contexts – social, political, historical, gender, ethnic' (Sim and Van Loon, 2000, p165). Critical theory is a tool which enables us to put engineering 'under the microscope' and to see whether we like what we see or whether we want to change it. For example, we might ask you to imagine that a large engineering firm is 'downsizing.' The management of the company need to consider the implications of what they are doing in order to make decisions. Workers are often faced with this situation and remain for many months or years unemployed after this event as they can find no work. This situation involves many complex arguments which we will explore later in the book. However, for now, to give an example of a critical perspective on this situation, we might explore the comparison between the individual versus social meaning of the situation. Osborne and Van Loon (1998) ask us to consider the question that some of these workers might ask, 'Why am I unemployed?' We might consider how much lack of respect an unemployed person has in society. Students of mine have actually commented that 'We can't help it if they don't respond to education.' It seems as if the common sense view is that the individual is responsible. In fact, there are many reasons for the surplus labour that causes people to be laid off which include:

Technological change (new machines).

Changed work practices (efficiency).

Work done in other countries (Globalisation).

Political change (government policy).

Cultural change (different products wanted).

Lack of requisite skills (no access to education or retraining).

None of these has anything to do with the individual worker but 'blaming the individual is common political practice' (Osborne and Van Loon, 1998, p10). Students will be asked to consider their own views and to debate and critique the various perspectives presented to them when focusing through 'windows' or lenses on typical areas well known to students. A very brief consideration of these windows will be presented below under the heading Lenses or Disciplinary Perspectives.

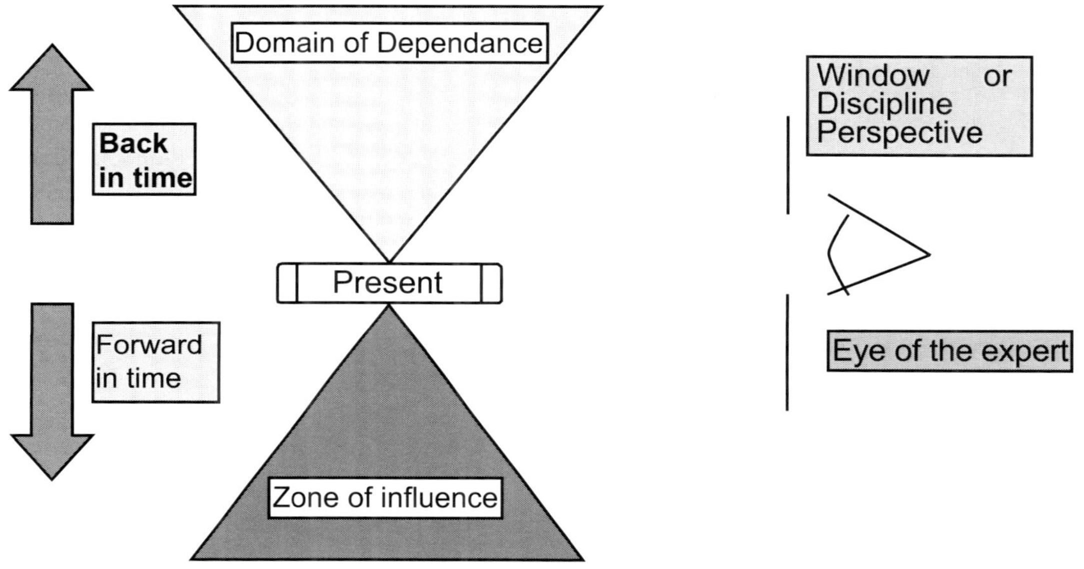

Figure 18.1. Windows of perception.

Perhaps then, we can extend the metaphor even further by placing various experts from different disciplines watching the unfolding of events through a window of a nearby building. Each expert will "see" the developing events within the context of a particular discipline or perspective. We will "see" that there are not events in history, sociology, political sciences, gender studies, economics, and global studies but simply events that can be understood in different ways and in different contexts. Each decision we make at this present moment in time is a result of the decisions and work of countless others at countless other earlier times. Equally as importantly, you will be introduced to the notion that each decision you make as a human being and as an engineer will have consequences that may extend far into the future and may be understood by others in a completely different way.

Let's reflect on the implications for our "zone of influence" and how we might relate it to our notion of "windows." The point we are trying to make is that what we see, what we come to regard as an objective truth, is a function of what we have experienced in our lives. An event which occurred during our childhood and how it effects our notion of reality is transported downstream much like the water that flows past the obstacle in its path. If this is true for events in our lives, then certainly, it makes sense to think of how our training and education can influence how we see the world, and this influence will continue to be with us as time advances. This idea can certainly be extended for the case of 'experts,' how they see the world and what implications these perceptions might have which are propagated forward in time.

Lenses or Disciplinary Perspectives

In the chapters that follow, we will not explicitly state each time we use a different lens, but to give you an idea of the different influences used to develop the thinking of this book, we present some of the typical lenses below.

Lessons from Sociology

This perspective frames how engineering is produced and utilized in local and global communities, how different engineering projects originate, how social needs are defined (participatory needs analysis), notions of social welfare and how the engineering impacts and affects different receptor societies. It also considers worker organization, employee relations, labour unions and social capital.

Lessons from History

This perspective deals with an exploration of the basics of the Industrial Revolution and the relationship between automation and labour. We consider the implications of this today with reference to globalisation and how it affects engineering organisations in developing countries. We will also explore the effect of increasing mechanisation on systems of thought and action.

Globalisation Debate

Current issues influenced by the global market will be considered using the work of key texts on globalisation - introducing the World Trade Organisation (WTO), 'free trade' global capitalism and sustainability.

Lesson from Economics

Different economic models will draw on a range of perspectives from capitalist, to Marxist viewpoints. Economic development will also be considered from the perspective of Amartya Sen's work on 'freedom.' His approach differentiates itself from traditional practical ethics and economic policy analysis, such as economic concentration on income and wealth (rather than on characteristics of human lives and substantive freedoms), utilitarian focus on mental satisfaction (rather than on creative discontent and constructive dissatisfaction) and the libertarian focus on procedures for liberty (with neglect of consequences of those procedures). There are, of course, connections to low income, but this needs to be integrated into a bigger picture. Poverty then can be seen as deprivation of basic capabilities rather than just low income, e.g., premature mortality, significant undernourishment, persistent morbidity and illiteracy.

Lessons from Political Studies

We will focus on the recent political changes that have occurred and the ways in which these affect the relationship between engineering and society. For example, the massive political-economic change that occurred with the collapse of state socialism, the establishment of private property, the influx of multinational capital into Eastern European countries and the integration of firms into the global economy.

Lessons from Environmental Studies

Recent developments on climate change, life cycle analysis, sustainable development, Kyoto protocol and different governmental perspectives will be explored and links to public health questioned.

Lessons from Psychology

From this perspective, we will consider personal development versus professional development. We will refer to recent work from psychology and Gestalt therapy, which helps the student consider their own blocks to self awareness, an understanding of personal ethics and potential ways of enhancing individual freedom and personal creative potential.

Throwing Away Rubbish

In this chapter, we are focusing on the basic service of rubbish or garbage collection. As a citizen in the Global North, we expect our rubbish to be collected on a certain day of the week. As an engineer, we might reflect on the processes of recycling, transport, collection and separation. Viewing this process from the perspective of social justice, causes us to ask different questions.

Figure 19.1. Throwing away rubbish.

What happens when you throw away your rubbish (or your trash or garbage, depending on where you are from)? Maybe, if you live in a very environmentally conscious city, you will have separated your rubbish into those elements which can be recycled and those which cannot. You may even separate out organic or compost material. If you do this well, you will notice how little rubbish can be left that is destined for the landfill. For some people, however, rubbish is not such a simple issue. For many people it is a way of living. This chapter takes you on a journey through the maze of complexity that is waste in Buenos Aires, Argentina where we worked for six months during 2007. It is the story of a project, Waste for Life, coordinated by Caroline Baillie and Eric Feinblatt, and the context in which they found themselves. Along the way, they learnt more about waste than they ever thought possible.

19.1   THE POLITICS OF GARBAGE, BUENOS AIRES, ARGENTINA

After the economic crisis in 2000, 80,000 cartoneros or 'cardboard pickers' were found throughout Buenos Aires city, collecting up to 66 tpd of plastics, paper and other recyclables. A new law on 'Integral Management of Solid Urban Waste' which the group Greenpeace helped draft went, into effect late 2005. This law, better known as 'Zero Garbage Law,' was additionally intended to bring about decent working conditions for the many informal garbage collectors or 'cartoneros.' The law stipulates that the amount of garbage in landfills is to be reduced by 50%, and to reach that goal, the Buenos Aires city government has sponsored the organisation of cooperatives of garbage scavengers, and provided space for the first warehouse [World Bank Development (1999), Gobierno de la Ciudad (1998), Aiello and Crajales (2001)]. However, there are many other disorganised and family run cartoneros teams that still exist. Currently, the estimate is anywhere between 5,000 and 20,000. It is still the case that over 90% of the city's recyclables are collected by the cartoneros, informal and unpaid workers who live in the shantytowns and enter the city centre by night to find the waste (CEAMSE, 2007).

Much of the recycling that is collected is sold directly to agents at a price of about 16c/kg. Some more organised cooperatives sort and sell the materials directly to industry, and others sort and reprocess, and in some cases, recycle. However, few create final products, and therefore, income generated is very low. The potential is high for composites made from upgraded plastic, particularly forms which have found no other market, such as plastic bags and some plastic containers; however, no such technology is currently utilised in the recycling circles of Buenos Aires. Fibres in the city that could be used to reinforce the plastic include cardboard waste and wood chips.

Waste-for-Life is a collection of projects led by our team in Canada and focussed on supporting low income cooperatives to enhance their income by helping them to develop natural fibre composites from waste plastic, and locally found fibre using simple technology – a machine called a hotpress – which presses and heats the materials. However, these cost upwards of $50,000 to buy off the shelf, and we decided to design a lower cost version. Darko Matovic of Queen's University came up with the design before we left for Buenos Aires. Waste for Life ran in Buenos Aires between July and December 2007. We first conducted a series of needs analysis interviews and group discussions with local stakeholders such as cartoneros, local government, landfill managers and workers cooperatives. Was the idea of making composites from waste a useful one? Did the groups collect enough plastic? How would they wash it? Did they have storage space? Did they have a source of fibre? What product would they make, and who could they sell it to? Would it be cost effective for them? What equipment would they use to process the composites, and how could they afford this?

When we decided that we had enough interest from local groups; we commissioned a hotpress - the machine was built by a designer in the Buenos Aires area.....

Instead of going into more detail about the project, we have decided to show you an abridged version of the blog that was created during the time of the project. It is possible to follow the team's progress and learn with them about the complex political and social issues that we had to understand and cope with. We have deliberately kept the flow of the blog, so that you can read along and live with us in the streets of Buenos Aires, asking the questions we asked and seeing if you would have made different choices. It is a living example of all of the theory we have been speaking about in the first two volumes of this book. Many of the blog entries appear not to be directly about the project, but these are critical for our understanding of the environmental, social and economic context, and in our assessment, of needs. We have, additionally, superimposed challenge questions, after the fact, for you to consider what you would have done and what you might do in the future if you embark upon a project in a context very different from your own.

The Blog. (Unless otherwise stated posts and photographs are by Eric Feinblatt and Caroline Baillie, coordinators of Waste-for-Life, Buenos Aires, and written in the first person regardless of who was writing.)

19.2   FINDING OURSELVES

June 5, 2007.

Arriving in San Telmo, BA, after a very tiring and annoying journey complete with delays and lost luggage, we opened the doors to our new apartment full of character, shuttered windows, and the charm of new discoveries. The neighborhood, with its cobbled streets, market places, cafes and restaurants, was just what we had hoped for. After dark, we went on a search for some food and there, right in front of our apartment, were a group of 'cartoneros' – a woman stacking a cart with cardboard, a young boy ripping open bin-liners full of mixed and rotten waste to locate any morsel of paper or plastic. Why were we shocked to see this? That's what we had come for, wasn't it? But somehow we weren't ready. Some part of us wanted to hold onto that feeling of the tired, annoyed traveler wanting sustenance, warmth and comfort. The shock of our ridiculous, late (not even fully lost) luggage, one-meal-a-little-late, hungry problems embarrassed us.

We are more than aware that many foreigners think they can come and 'fix' problems in contexts other than their own. We do not wish to repeat the errors of others and hope for advice and partners who might help us with this project. The questions we thought about before we came keep reasserting themselves: Who should we look to partner with? What is the government's role in waste management and collection? Could our project possibly perpetuate a system of inequities? How will the proceeds from the materials and products the groups collect/make get distributed? Are we helping to launch an economic model that we ultimately do not support? How do we avoid profiteering from outside agencies as soon as there is a profit to be made? Who supports us? What affiliations do we have and how do they influence our values and our commitments? Is any of this our business anyway? Should we just stay at home?

Then there are the questions of the photographs and the website. What right do we have to upload photos of people whose life circumstances and choices we know little about? What is the purpose of our website anyway? To give ourselves a pat on the back that we are actually doing a project and progressing – even if progress is measured by images and chat? The cartoneros have a very public life as it is. They are on view every evening as we see them collecting their source of income. Furthermore, this is not just a Buenos Aires issue. Living off waste is a universal phenomena, complicated by governments who clean up only the nice parts of town and who recycle when it suits the economy. Before we left, we saw a scene that almost exactly replicated the scattered waste seen in San Telmo, traumatizing the pavements of our Bronx home.

Figure 19.2. Rubbish in Buenos Aires.

Figure 19.3. Rubbish in the Bronx, NY.

Challenge Box: What are your thoughts about what you have read so far? What right have we to come to Buenos Aires and think we can help? Locals can be quite upset when they think we are here because it is a city in need of development. Most portenos (people from BA) believe Argentina to be 'first' world.

19.3   KINGSTON HOT PRESS

Published by
Darko Matovic at July 17, 2007.

Here at Queen's University in Kingston, we are making the first press prototype these days (July 2007). The whole process is quite exciting for us all, but sometimes frustrating, of course. In the pages below, I will describe the press design, point out the things to watch out for while building the prototype, and describe the building process in a step-by-step fashion. As we gain experience testing and commissioning this prototype, I am sure that there will be changes necessary, but, of course, we are not aware of them now. As usual, every prototype is perfect on paper and on the screen, until built and tested.

The Kingston Hot Press (KHP) is a manually operated hot press designed to make tiles of composite plastic, up to 610×610 mm (24×24 in) size, 1-10 mm thick. The press is designed to provide up to 6 MPa pressure (870 psi). The total force required for pressing is thus around 200 tons (2 MN). The temperature can be adjusted up to 250°C. The key design challenge was to make a manually operated press able to supply very large force, and yet be affordable and relatively easy to make.

Here is what the current press design looks like:

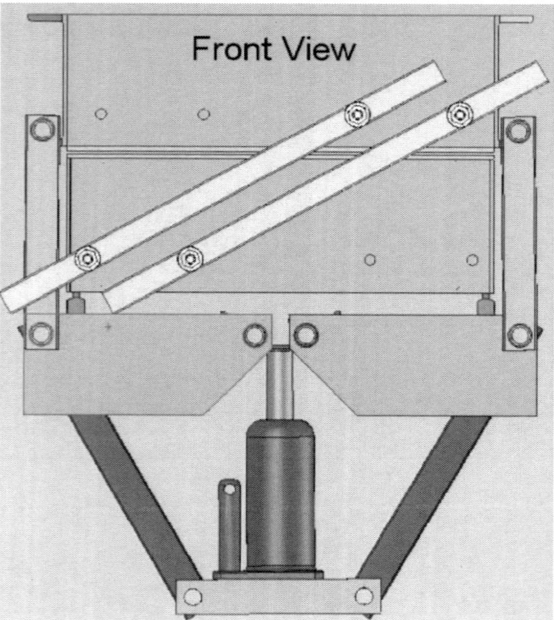

Figure 19.4. Design of the Kingston hotpress.

19.4 THE PRESS IN THE MAKING

Published by
Darko Matovic
July 18, 2007.

My current posts come from Kingston, Canada, where I teach at Queen's University. This is where the idea of designing and building the hot press came, after hearing about the efforts to startup composites production in (another Waste for Life) location, Lesotho, using local fiber and recycled plastic.

The hot press is essentially made of three functional units: (1) the lids, (2) the lifting mechanism, and (3) the stand. The "core press," consisting of the first two units is shown below in the triaxial view:

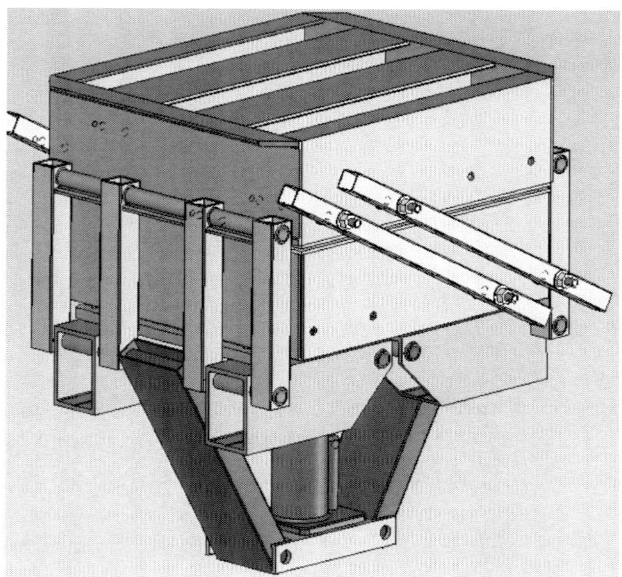

Figure 19.5. Alternative view of Kingston hotpress.

The top and the bottom lid are similar. Both are welded frames, each made of four "C" channels and two "I" (or rather "H") beams. Each frame supports a square plate, two plates pressing against each other, squeezing the tile in between. Inside the frame, there are three "U" shaped electric heaters in each lid, attached to the back of the plate, in between the frame "ribs," i.e., the steel profiles. The steel profiles are BIG, 200 mm high (8 in). This is what you need when trying to press a 0.6×0.6 m area with 200 tons, and not allow for more than 0.5 mm bulging in the middle. At least, this is what the finite element simulation says. We'll soon discover if this is right, not enough, or an overkill. This is what prototypes are for.

The top and bottom lid are similar but not identical. The top one has two C channels on the opposite side oriented outward, while the bottom has all four C channels pointing inward, mitred and welded as a picture frame. Here's the top frame in the making, photo taken today:

Figure 19.6. Making the press in Kingston.

19.5   WADING INTO THE WATERS

July 20, 2007.

We left North America two weeks ago with a handful of contacts in Buenos Aires, a sample 3inch x 3inch composite tile fabricated in Caroline's Queen's U. Lab, years of experience as educational developers, scientific knowledge about waste plastics and natural fiber composites, and a desire to share what we know with self-managed groups interested in developing poverty-reducing solutions to a specific ecological problem in BA. We presumed that our project would revolve around one or more of the cartonero collectives because they had access to the composite raw materials – the waste plastic bags and cardboard which are the material centerpiece of Waste-for-Life Buenos Aires – and because they are visible, often well-organized, but an at-risk population.

To the world outside of Argentina, El Ceibo, which is made up of 50+ families, is the most recognizable cartonero collective. It has been studied by university researchers and featured in the popular press, probably because of its tantalizing collaboration with residents of the up-scale Palermo neighborhood in BA. Cristina Lescano, it's founder, has more than 20 years experience as a successful community organizer, and we had been in touch with her through her 'secretary' Jim McAsey, a New York community organizer who had taken a year off to come live in Buenos Aires, inspired by the city's various self-managed collectives and, in particular, the 'recuperated' factory movement.

The waste plastic bags that the cartoneros collect have no value to them because they cannot be sold to recyclers. They end up in landfills or dumped in riverbeds on the outskirts of BA. Christine zeroed in on the commercial possibilities of 'upgrading' the plastic with cardboard fiber to create

Figure 19.7. The composite tile made from plastic and cardboard.

Figure 19.8. Cristina Lescano.

some sort of marketable building material. Of course, this would dramatically change the nature of El Ceibo's work from collectors and re-sellers of waste to manufacturers and distributors of finished goods. Such a focus shift raises all sorts of questions for El Ceibo and ourselves that we will be addressing in later posts.

19.6   INSPIRATION

July 22, 2007.

Last week, we also met with Dante Munoz and Carlos Levinton from the Faculty of Architecture at the University of Buenos Aires. Carlos directs the Centro Experimental de la Production (CEP) - Experimental Recycling Centre - and welcomed us as family members into its fold. We were introduced to the fabulous projects that they organize, including what they call social factories, working in poor areas, where they teach children new technologies so they can teach their parents. They have embraced all sorts of exciting recycling projects and educational experiments as well as the development of a recycling museum, now sadly closed. Dante and Carlos invited us to take part in a class for teenage school children one morning, and we really started to understand what their group was about. They are innovative, caring, socially minded, working with all manner of groups in different areas of Argentina and Bolivia. They function on the voluntary support of faculty and architects, as well as locals, who wish to get involved. It is truly inspiring to see this group in action. We were very moved by the commitment and love that is shared and that fuels this enterprise.

Figure 19.9. Carlos' group at work and one of his student workshops.

19.7   THE WORKERS' ECONOMY

July 23, 2007.

Over the last few days, we have been completely exhausted, stimulated, moved and frustrated by the conference 'The Workers' Economy: Self management and the Distribution of Wealth.' It is not news to the progressive that something interesting has been happening down here in Buenos Aires with the 'recovered factories' - those enterprises that were taken over by their workers when they went bankrupt in the economic crisis of 2001. The conference was hosted by the College of Philosophy and Literature at the University of Buenos Aires, and we were kindly invited by Marcelo Vieta, co-host and fellow Canadian academic originally from Argentina, to make a presentation about our project. We were now really entering very new territory, and it was with quite some trepidation that we entered the Spanish speaking world of commingled academics and workers from the recovered factories. Albeit wonderfully translated by a team of extremely patient experts, it was sometimes hard to concentrate on the endless 12 hour stream of papers that had been produced all on different facets of the same few intriguing factories. What made the meeting so memorable was, however, the commitment to the idea that here was history in the making, here was hope, here was imagination, here was a future that did not involve competition and selfishness but sharing and values which might be worth knowing. The workers themselves were the most profound when they spoke about their take-overs, their families, their problems and their dreams that they were really doing something worthwhile with their lives, that they could never have imagined possible. As we sat and listened to the representative from Hotel Bauen, the recovered hotel (which, despite its financial issues, was hosting those who could not pay during their stay in the city), a notice was posted that the Hotel workers were going to be evicted in a couple of weeks.

Figure 19.10. Bauen eviction notice.

All these hard working men and women, from all manner of organisations, told us that all they wanted to do was work. In order to break even, with all the accumulated debt of the owners, they were paying themselves pitifully small wages but were proud to be producing and serving. A highlight for me was the evening performance of theatre in the recovered factory of Chilavert printers.

Figure 19.11. Chilavert Theatre.

Community, culture and sharing is commonplace around these people. Students were flocking down to study this space, this phenomenon, and we were privileged to be among those who worked and studied in an area which seems to put flesh onto the starving bones of theory. On a personal level, we wanted to make contacts to see if any of the skilled workers in the factories might be keen to make our first hot press. We also met many academics who share common values and who we hope will help us understand the socio-economic-political implications of what we want to do.

Challenge Box: Consider the plight of the Recovered Workers. What are your thoughts about what they have done? How does this compare with workers in your own community?

19.8 THE PRESENCE OF ABSENT ONES

July 30, 2007.

Previously, we mentioned the friendly welcome we received from Carlos Levinton and his research group at the University of Buenos Aires. We'll write more about what promises to be a very fruitful collaboration with them in later posts, but want to point back to an unsettling experience for us when we first entered the School of Architecture two weeks ago.

Figure 19.12. Banner in the Faculty of Architecture of the 'Disappeared.'

Hanging above the lobby of the Faculty of Architecture is a stark black and white banner of young faces. They are pictures of students and faculty who were murdered and 'disappeared' during the dirty war of the late 70's and early 80's, and their faces are a grim reminder of the price of political opposition. Argentina lost a generation of university activists whose ranks are only now beginning to be filled by a cadre of incoming students, some of whom we were fortunate enough to meet shortly after our arrival here. We, in North America, have not experienced similar repressions or resistances, although we did have our Kent State, our police riot during the Democratic National Convention in Chicago, and mass university mobilizations against the Vietnam War. (I am aware that this sentence excludes the day-to-day experiences of many U.S. minority groups.) These have all but faded from our collective memories, and the notion of the university as a site of learning and engagement seems peculiarly anachronistic. In our talk at the Workers' Economy Conference, we spoke about the university and its original linkages to an educated citizenry who could maintain a democratic and just society. This is one of the ideas that fuels our work here in Argentina and that is continually being reinforced by the people we meet and the work that they are doing. It is an idea worth remembering and speaking about, as are the faces on the banner which simply says: 'siempre presentes,' always present.

> **Challenge Box:** Find out more about what happened in the Military dictatorship in Argentina and just how many people disappeared for what they believed in. These questions as with understanding the recovered workers - help us to place ourselves in the context we are working.

19.9   LOST PLACES

July 30, 2007.

Ok, so I thought I knew how to use the 'Guia' – the incredible bus route map for the whole of the city. I was getting impatient and decided that we had to meet with some more cartoneros. We had contacts and I decided to call them up despite my horrible Spanish. After the call, I partially believed I had made an arrangement for 3pm and I thought I knew the address. The bus, however, had a different idea and decided to go west at a very crucial point. Another bus and a long walk later, we were in an extremely poor area, with private police seemingly patrolling a kids' football game. Large cameras and American/British accents were bad ideas. I was trying very hard to locate an address which appeared not to exist, amongst many stray dogs (and I didn't have that $600 rabies jab) and rubbish dumps. Finally, I asked a woman who didn't really know so I called the host and he explained to her and a growing throng of local kindly souls.

Several phone calls later we were in a small and very unreliable van driving over lots of holes to a totally different area of the 'favela.' We were told on the way that this whole area was supported by collecting rubbish. We were at the heart of the Dock Sur cartoneros patch. Our kind saviours deposited us at the right location, home of Carlos Perlini's Cooperativa de Trabajo Avellaneda Limitada. We were shown around the Galpon (warehouse) which included bottle stripping and weaving with a basic weaving machine. They also had equipment that could have, were they to have the funds to finish the machine, chop up plastic crates into small pieces which could be sold at a much higher price. Carlos told us they bought plastic waste from the local cartoneros and sold it - having washed it, dried it and packaged it. If they were to have the funds to support equipment they could create a greater value for the plastic they sold. This small, ridiculously under-funded facility supported 50 families.

19.10 SEPARATION AND SOLIDARITY

August 2, 2007.

Today, we went to the Bajo Flores recycling centre in another very poor barrio. The collective here has developed a separation plant - the first of four planned in Buenos Aires, where the members buy waste, delivered to them in trucks (which sometimes arrive at their destination), and separate it manually with the aid of a large conveyor belt system which enables them to separate plastic from glass and bottles from bags. The system is a standard design, which I have seen operating in the UK and Canada, and waste is further squashed into large cube shaped bales for selling on to the next in the chain. The equipment was apparently bought for the group by the city government, who also provided the warehouse. This is an interesting system. The city government, for the price of a warehouse and two pieces of equipment, gets much waste processed. No labour cost - as it is self run, except for a small subsidy to keep the peace. Is this the government shirking their responsibility? In the UK, U.S., and Canada, we expect the government to pay the salaries of our recycling workers. However, members of these self- run units are, seemingly, content to be their own bosses and fear that changes in the administration may mean they lose their livelihoods.

The most amazing thing for me is that people here seem to work together so well. We noticed this when we met the recovered workers. We were told that this was because they have become used to a lack of administrative support over so many years of loss and betrayal, that they have had to learn to support each other. I'm not sure if this can be the only reason, but we have felt privileged to work amongst these committed teams. Our host today, Silvia Rossi, another member of the University CEP, is, like many of the group, a self employed architect who works with the centre pro bono one day a week, and does many projects like this one in Bajo Flores on her own time. It was she who persuaded them to get the warehouse and equipment to get themselves started. She tirelessly supports them on a weekly basis. I will bring more than memories back from this project.

> **Challenge Box:** Who do you think should be responsible for collecting and recycling rubbish? What do you think about the government using the cartoneros to do the recycling? In some cities this is even more well established such as Curitiba in Brazil.

19.11 WHO OWNS THE WASTE?

August 2, 2007.

The question of who owns the waste is a constant theme here in Buenos Aires. It is clear that the residents are made aware that they don't own it once its put in the street. The private/public organisations that run the trucks to collect the waste believe they own it, but their interest is not in helping the environment - they are paid to deliver to the landfill. Some of the workers break open the rubbish bags to relieve them of the recycling inside thereby adding many hours onto their roadside collections, but increasing their income by independently selling cardboard and bottles. The cartoneros who pick their way through the bags left on the pavements late at night, know they don't own the waste, but the tendency to make something from the discarded bodies and belongings has apparently been around for 200 years since the 'Ciruja' - when animals bones were made into useful objects. Some groups have evidently made arrangements that we don't fully understand, as they have access to the waste in whole districts without having to 'waste-pick.' Some are legal and some not, this is clear. But who does own the detritus, who do I give it to when I throw it away? I've given up the right, so who takes it? We have become so accustomed to idea that the government will take it away and that we pay them to do that through our taxes, so its a dirty and necessary job. We believe that putting our bottles in a separate bin means we are saving the planet, not realizing that much of this waste will be sold to a developing country that has no recycling system of its own. Their own waste then goes into landfill. The self employed, self organised cartoneros are doing the city's dirty work. They are amongst the few who are currently preventing the whole lot from going to the landfill. And they might earn a few pesos a day if they are lucky.

19.12 VARIATIONS

August 15, 2007.

Every time we leave the city we travel through the most incredible deprivation, the villas miseries or shantytowns. Houses are basic constructions, made of blocks and tin roofs, and many of these areas have no electricity or water and sewage services. 100,000 live off waste in the province of Buenos Aires, and many of the cartoneros travel from these areas into the centre of town to collect the city's detritus.

A visit to Maria Virginia of Abuela Naturaleza on Saturday proved to be extremely useful. She is an incredible 'entrepreneur' who has decided to spend her life working with waste recycling. She has been involved with this for 20 years, many years before the excitement about global warming. She teaches children in schools and fetes about recycling using puppets and other creative techniques and has at one time been involved with a cartoneros' cooperative. What she is doing right now, however, is providing a huge source of information and encouragement about recycling whilst making herself a living. She has organized her local community to separate products that can be recycled and goes around in a van to collect it. Although her van is at the moment being mended, she still receives visits from her neighbours with their deposits of recycling. Virginia has found a market for practically

Figure 19.13. Maria Virginia and her waste sorting depot.

everything she collects and showed us the huge number of items, all separated with the help of local workers that she employs. This is the most organized collection of recycled goods I have ever seen. Furthermore, she receives for most items about 1.5 pesos per kilo. Not much, but when you consider that the cartoneros receive about 20 cents per kilo for much of their produce, sold to agents who may sell it directly overseas, we learnt that here was someone who had found a way to actually recycle material and make a decent living. She hopes to pass on her knowledge to cartoneros' groups in the future. We also learnt about those materials that she cannot, at the moment, find a market for. These are a range of items made from plastic, cotton and cardboard, ideal for processing with our hotpress into composites. She even suggested a possible product for the materials.

19.13 BOTTLES ON CARS

August 15, 2007.

Figure 19.14. Bottles on cars.

We've been in BA for six weeks and during each of those weeks we've seen parked cars or pickups with large liquid-filled bottles on their hoods or roofs. At first, it was just a curiosity, and we had fun trying to guess what was going on, but eventually the persistence of the sight made it clear to us that this was a custom, not an anomaly, and that we didn't have a clue about the meaning of what we were looking at. This has been our experience with many things here.

Late the other night, we saw three CLIBA (Compania Latinoamerica de Ingenieria Basica Ambiental) garbage men rip open the large green plastic bags that they had thrown into their truck, sort the recyclables, and put them into their own 'for glass' and 'for can' bags - presumably to sell off privately. CLIBA is one of several private trucking companies that the city hires to pick up city waste and haul it to one of four CEAMSE landfills where it is sold by tonnage. CEAMSE (Coordinación Ecológica Area Metropolitana Sociedad del Estado) is the large public private organisation who owns all the landfills. This is a system that has many interconnecting players in it, each of whom claim a certain proprietary ownership of the garbage - for garbage, we are learning, is a very, very, big business here in BA. Four years ago, Mauricio Macri, the man who recently won the Mayoral election in BA, and who has a considerable commercial interest in garbage removal, accused the cartoneros of thievery because they were 'stealing his garbage.'

Figure 19.15. Garbage workers in BA (faces blanked out as they had been committing a 'crime' taking recycling from the garbage).

> **Challenge Box:** What do you know about where your rubbish and recycling goes to?

We spent part of a day last week at the Bajo Flores Green Point, the first of six projected model centers that the municipal government and foreign investors are funding around the city. Each of these centers is run by and for the benefit of a different local cooperative - often a cartonero collective - and is supplied waste to sort and recycle by one of the private garbage hauling companies. Bajo Flores is supposed to receive garbage from some of the city's five-Star Hotels and apartment buildings over 19 stories, but it was pretty obvious to us that the center was working way below capacity. What was happening to the garbage along the route? Though destined to Bajo Flores and the 40+ families who were working there, we learned later that the truckers or the hotel employees or the apartment building janitors were diverting it, and someone (who knows who?) was selling the recyclables on their own. There are lots of competitors for garbage and many people could profit making certain that Bajo Flores fails.

Figure 19.16. Workers at Bajo Flores.

Last week disappeared very quickly in a kaleidoscope of visits to all manner of extraordinary places outside the city. One visit was to La Vallol we where went to visit a 'Barrios de Pie' home building project. Except, there were no building materials and so no work was actually going on. Apparently, the money/materials from the government had not arrived, so empty shells of half-built houses lay dormant. Barrios de Pie is a politically astute and powerful association of workers' cooperatives across the country. Because they have an internal economy of sorts - one cooperative selling its goods or services to others - it is an appealing organisation for us to work with. There would be no need to identify or create an outside market for the composite products. For instance, the Barrios de Pie collective that picks waste plastic from the riverbeds, could clean it and sell it to the materials fabrication cooperative, which could process it into building tiles, which could be sold to the housing construction cooperative. And the hotpress that is needed to process the plastic could be manufactured by the machinists' cooperative. This would be a tight scenario where the beneficiaries of the composite product or products could be easily identifiable throughout the supply chain. Dante, who works closely with Barrios de Pie, took us to La Vallol, where the housing construction cooperative is working on building the first 16 of 180 projected homes. We wanted to look at the housing construction to see if there was any potential need for the composite tiles, but nothing was going on at the site - there was really no one to talk with. A few homes - partially finished - stood, and a few foundations were poured and ready for construction, but there were no workers around to work. The housing cooperative, which had won a contract from the government to build the homes, hadn't been paid the money needed to buy the materials, so work was at a standstill. Were we simply looking at a bureaucratic snafu?

The bottles? Oh yes, they mean that the vehicle is for sale.

19.14 THE PLOT THICKENS

August 16, 2007.

A meeting with the city government office, specifically the research group of the Politicas de Reciclado Urbano, proved quite a remarkable surprise. From what we had heard both before and after our arrival, we imagined a meeting with some dusty, defensive bureaucrats who would avoid all controversies and present their picture of success with academic splendour. Not a bit. Here was a group of four young researchers who really seemed to care about what they were doing - not just for the environment, but for the people. We can't say they represent the whole plan or the official view of the government, but they were a breath of fresh air. They were assigned to research the issues of waste in the city and told us much about the recycling centres, confirming the stories of the lack of success, non delivery of recycled materials and general confusion about where this waste does go. They also could tell us which reports had been written by whom and whether their studies on waste collection and recycling referred to both formal and informal sectors, domestic, street waste and other waste or just formal domestic. They estimated 11% of the city's waste was currently recycled, and that most of this was currently carried out by the city's 4000 cartoneros, along with El Ceibo as a major player. Five cooperatives have been singled out to run the recycling centres, as we had been told before, because they were the most organised. However, this was also the office, which registered the singles and families. So long as they were older than 15, we were told there were no other restrictions, this was not what we had been told before. They were given gloves and a safety strip to wear over their clothing, and so long as they did not disobey two basic rules, they could legally work in the streets. One rule was that when they ransacked the bags of rubbish they found there - that they would not leave the rubbish all over the street afterwards. And the other rule was that they would not stack up cardboard and plastic for their van so they could pick it up later. These two requirements are quite obviously not being obeyed by many groups as you see when you walk home at night, avoiding the spills of tomatoes that were strewn across the pavement. Furthermore, at many street corners on certain nights of the week, you would find a large pile of sorted recycling materials. Maria, Antonella, Mariela, and Felix were incredibly friendly, very concerned about the cartoneros and what might become of them after the forthcoming city government changes. It was difficult to doubt the sincerity of this group. We discovered that Mariela was a radio DJ for the Bajo Flores radio station. She told us she had to do something to feel alive and she went there as often as she could. The plot thickens every day.

> **Challenge Box:** Before we came to BA, we had been warned that it might prove difficult for us if we were seen to be affiliated with any group - government, private or social collective. So far, we had experienced a wary welcome by everyone, which turned into great friendship on only the second or third visit. Do you think it was the right thing to do - to remain unaffiliated? Why or why not?

19.15 SOFT COLLABORATION

August 17, 2007.

INTI (Instituto Nacional de Technologia Industrial), a federal government institution here in Buenos Aires, has some very intriguing functions. I would have imagined that it might be able to help us identify the location of a hotpress or advise us on other technical matters but, in fact, we stumbled across Hector Gonzalez at the Workers Economy conference and, finally, went to see him today. His job involves coordinating research, which supports the local cooperatives. Whatever comes in – they try to farm out to various of their departments. These are social, economic and technical issues. Instead of being greeted with a discouraging what-can-we-do-for-you smile – we were invited to meet others from architecture and construction. They had already downloaded our blog and were ready and willing to collaborate and help in any way they could. Our mission to help cartoneros seek extra income was also theirs, it seemed. We have now found quite a few left leaning government employees, all of whom are a little nervous about incoming city governments and the changes they will make. We were offered assistance to find the standards which will make our materials legal in buildings and in furniture, help in locating a hotpress or personnel to make one, the names of recovered factories who might make the press and finally offers of 'soft' collaboration.

19.16 NO COST HOUSING

September 3, 2007.

Shortly after arriving in Buenos Aires two months ago, we met up with Dante Munoz and Carlos Levinton. They have tutored us in the politics of waste and the re-purposing of waste (as well as ceviche and milongas), and we have met with them at least once a week in the Centro Experimental de la Production (CEP) at the University of Buenos Aires' School of Architecture, Design and Urbanism. Carlos is an architect, a professor of architecture and director of the CEP, which is supported by (mostly) volunteer architects and funded by a three-year, $7000/year grant from the University. The CEP has been described to us, and probably to University Administrators, as a research unit that investigates emergency disaster preparedness and response (like a FEMA think tank), but we are learning that it is much, much more than that and, in a conversation with Dante the other day, we began to get the bigger picture.

When the CEP thinks about minimizing risks to disaster, they are thinking about theoretical models and practical applications that reduce vulnerability to environmental, economic, social and natural disasters, and their work attempts to address some very fundamental questions such as: what is healthy housing? What are healthy cities? What are healthy exchange relationships? What impedes health, and what political, economic and educational activities are necessary to restore it? They are part of a larger, loosely joined movement of academics, collectives, and activists who, perhaps, inspired by the self-management movement that has flourished since the 2001 economic collapse of the Argentinean economy, or Argentina's history of legal and successful workers' cooperatives, or the scattered but concentrated experiments with Trueque (the Inca word for a multi-reciprocal exchange economy), believe that they can mobilize against the values of a consumer economy that processes identity and relationships in terms of acquisitiveness, accumulation and privilege. We share many things with the CEP, not the least of which are our mutual experiments in adding value to what is normally considered society's waste. We are working with low-density plastic (plastic bags) combined with natural fibers (cardboard) to create building materials and domestic products, and they are producing (in addition to many other things) solar heating panels and insulation from discarded 2-liter cola bottles. Both of us are investigating the use of low threshold/high impact technologies to improve peoples' physical living conditions and to reduce the proliferation of waste.

19.17 WATER TURNS TO MUD

September 3, 2007.

We have now been here for two months and I have the feeling that we are no longer wading into waters but though thick mud. I sometimes have no idea how to move or where to go next. I think this is an important feeling and I need to understand better what it means. Even speaking the same language in Canada (being British), I have the idea, sometimes, that I really don't 'get it.' Here with the language and so many different influences, historically, politically and economically, I realise how hard it is to work in another context.

The last week saw many meetings again, hopefully, leading somewhere but to where I am unclear. Also, we were joined by my cousin Karen's daughter, Emily, who is a geographer, and enormously helpful with the questions she is asking. We met with Brendan from Working World, who shared with us some fascinating insights about their incredible micro funding organisation and their views on the cooperative structures here. We also visited a very poor barrio in Moreno, where Carlos Levinton was teaching school children how to make hot water from waste plastic bottles - we even saw one of these 'machines' in action in a local house - the water was incredibly warm.

It was great to see how the work the CEP does manifests itself with the children teaching their parents. I had fun teaching some words of English to some of the children. They didn't understand why I spoke so strangely and when their teacher explained that I was English, they wanted a lesson. We exchanged words, naming bracelets and body parts and animals. I learnt as much as they, which was (to them) a surprise and much fun for all.

Figure 19.17. Learning English.

Our main focus this last week, however, was to visit two of the groups that we hope to work with. A return visit to Bajo Flores sorting unit, to discuss the next stages, and to make plans for a theatre group in their community centre (a sideline that Eric and I want to pursue here) and the most exciting visit - to UST (Worker's Solidarity Association) in Villa Dominico. We were hosted by our friend Marcelo and met Mario and his son who explained how UST works. It seems to be a 'recovered' enterprise in the sense that it extracted itself from the withdrawing corporation TechInt, and negotiated with CEAMSE to become contract workers, to maintain the regenerated land on top of a huge former landfill. They also work with seven other cooperatives amounting to 80 people who work in a variety of areas such as recycling and building. Furthermore, they told us have the

Figure 19.18. A solar water heater.

skills to make the machine. Mario was enthusiastic and immediately gave the hotpress plans to his colleagues to start thinking about the design.

19.18 TRANSFERRING OWNERSHIP

September 4, 2007.

Today we took the city office up on their offer to 'come up anytime the light is on' and asked more questions of Mariela, who is truly becoming our heroine. She knows so much and cares even more. I was interested to hear that cartoneros were commonly considered 'good' if they formed cooperatives, but that the chances of all cartoneros forming co-ops was extremely unlikely. We raced from there to visit Marcela de Luca, an engineering professor who works for both University of Buenos Aires and as a consultant funded by many different organisations including CEAMSE. We knew she had studied cartoneros and waste management in the city for years but had no idea what

Figure 19.19. UST Cooperative.

a wealth of information she would be. Again, she really cares about the cartoneros, but in a very different way to others we have spoken to. Marcela worries about the under 15 year olds who work alongside their parents and do not attend school. She gave us a figure of 9,000 registered cartoneros, higher than we had heard before. She also informed us that they had not really existed before 2001 - at least in their current status - different from the notion of ancient ceruja influences. Apparently, they were at work in the city area but not in the outer districts, and their functions were very different. They would collect directly from city offices and not so much from the streets. Marcela has created many, many reports and is extremely knowledgeable so we could only capture a tiny amount, but took away 6 tons of reports to read later (thank God for CDs). We also learned more about CEAMSE, which has become an enigma to us (even the city government say they don't actually understand how it works). It was started by the military government in 1978, seems to be a complex amalgam of city, provincial and national (maybe) government and private companies. They own the landfills, and as soon as the trucking companies drop off the rubbish that they have collected from the streets at transfer units, they own the rubbish. Until then, the municipality owns it. Are we starting to see the light? Not really. Who controls CEAMSE and who benefits and how privatisation works in this instance is intriguing to say the least.

Challenge Box: We have spent quite a bit of time trying to understand the context here in BA. Before I came, I was told by one well meaning social scientist 'stick to the science.' As you know (from Volume 1) there is no such thing as science in a vacuum. But how much did we need to know, and how did we know when we knew enough?

19.19 SORTING OUT THE DIFFERENCE

September 9, 2007.

A visit to CEAMSE's landfill, 'Norte 111' involved a tour around three sorting plants. The evolution of the tour is in itself a fascinating psychological journey. We got into a small air conditioned bus with some business men in white shirts and a very different mission and lived the experience of globalization within 30 minutes. The first visit was to the Social Factory on CEAMSE's property - run and staffed by cooperatives' members from the local barrio. CEAMSE gives them the waste, they separate it with rudimentary equipment and sell it, in order to pay themselves. One worker told us how lucky they were to have been selected to work in this place. The next location was a private factory adjacent to CEAMSE, which had more impressive equipment, and workers there told us that the compressor (to squash plastic and card into cubes to sell on) was the only one of its kind in South America. The final location was a Chinese run factory currently being established. We were shown into a huge warehouse, with machines resembling giant green dinosaurs waiting to be fed the waste that would come in tons. This was the largest scale I have ever seen. One of the businessmen suggested that the Chinese equipment was in fact old fashioned and would not work. He preferred a completely labourless, automated factory. Cartoneros watch out – your time is limited I fear.

A visit to CEAMSE does not necessarily include a visit to the landfill. As we were not official visitors, we had to use much persuasive power to get to the landfill itself. What CEAMSE did not want publicized is that the cartoneros still come in the hundreds every afternoon. They think that showing it on films and TV has increased the number of cartoneros that come in the hope of finding enough waste to live for the next day. At three thirty, they start to arrive - all ages – playing football while they wait, on their bicycles at the ready. When five thirty comes around, they are ready to race on their bikes the final two km to find the best of the recently dumped waste. Meantime, the animal-like machines wait in line to drop off their rubbish onto the open pit for the scavengers to explore…

19.20 REPROCESSING COOPERATIVES

September 9, 2007.

Two cooperatives funded by Working World, Etilplast and Villa Angelica, are examples of small family co-op organisations which buy waste plastic and reprocess it to add value to the chain. Villa Angelica has a yard full of waste and two machines, which chop up the plastic. When we visited, we were told that they had recently lost their only client and were hoping to mend their small extruder to seek new buyers. All the equipment in the yard was priced in terms of the number of tons of plastic waste that they had traded to get them. All these machines were worth less than a few thousand dollars. A more successful organisation, Etilplast, had an enormous extruder, which they had built themselves over some eight months. All machines there were huge, and according to Working World, a little over the top - industrial size machines rather than cottage industry. From our perspective, this group seemed as if it could make the hotpress, but how complicated they would make it and how long they would take, one could only guess.

Figure 19.20. Etilplast cooperative.

19.21 PRESSING ON

September 9, 2007.

Carlos Perini and his cooperative, Avellenada, was in a much healthier place this week when we visited. I wanted to make sure Carlos did not think we had forgotten him even though we still didn't have any firm plans for a hotpress, so we returned and met with several members of the co-op. In the last few weeks, they have stepped up their operation and have joined with another cooperative that has brought into the marriage a working plastics chopping machine. They can now collect, hand sort, chop, wash and dry plastic for sale by type. They had also discussed our project amongst all the members and were extremely keen to go ahead – assuring us that they have the skills to make the press and that all that is required is the funding to support the materials, the hotpress plans and training to make the composites. We have sent them the basic instructions to start things moving and wait avidly for the final plans of our press from Canada.

19.22 URBAN RECOVERERS

September 11, 2007.

Yesterday, Maria Virginia welcomed us again into her home in Ituzaingo. I am not sure why it surprises me that life moves on so quickly for our Urban Recoverer friends. Virginia had mended her van and was about to sell all her current supplies of recyclables, as she had to move things from her borrowed store (the next door neighbour's house) to her home. She clarified something that has been bothering me. Why was it that some of the cooperatives that we had met did not call themselves cartoneros when they clearly collected recycling materials? Virginia told us that the term 'cartoneros' really only refers to the very informal gathering that happens in the streets and, of course, to the most popular cardboard that they collect. The term Urban Recoverers is the official name preferred by those who collect from door to door and have a slightly more stable, organized existence. We had a very useful discussion to take our project on to the next stages and, like Carlos Perini, she is very keen to move ahead if possible. She thinks she has a contact who could build the press, she has a proposed product (recycling bins) and much waste which does not currently have a market.

19.23 BAJO FLORES

September 25, 2007.

Bajo Flores is one of six sorting centers or 'green points' that figure prominently in BA's Zero Garbage Law plans. We'll speak in detail about the significance of these centers in a subsequent post, but mention Bajo Flores here because it is where we have been running our weekly theatre workshops and because, in an earlier post, we noted that it was obvious to us that it was not working anywhere near its capacity. We went by yesterday for our normal 3 o'clock Monday rendezvous, and were told that it was cancelled because there was simply too much sorting being done. (Normally, the workers who participate in our theatre workshops are let off work early.) I walked into the sorting area and took some photos and videos of a very, very busy workplace.
What happened?

Bajo Flores was supposed to receive the raw garbage of BA's five-star hotels and some of its over-19-story apartment buildings. Two of the five private trucking companies and the single government-run company that are responsible for cleaning the streets and hauling the city's garbage were charged with bringing the waste to Bajo Flores to be sorted. The cooperative members separate the different types of plastic, the bottles, the carton, the paper, etc., from the other garbage - that is then hauled to the CEAMSE landfill - and sold to the industries that recycle it into paper and other plastic and glass products. In theory, it's a good system: the cooperative is responsible for sorting and selling, and it bypasses all of the middlemen that stood between its members and the final destination of the recyclables when they were working informally as cartoneros on the streets. The members share the profits equally; less garbage goes to the landfill, and the plastic and bottles and cardboard are reused. But it wasn't working, we were told, because the recyclables disappeared somewhere along the routes to Bajo Flores, and the cooperative (Bajo Flores Ecological Cooperative of Recyclers) was left with little to sort or sell. We learned, today, that the government decided to flex its muscles a little (it was their plan after all that was going awry) and called on their Secretaria de Inteligencia to get involved. They used GPS technology and planted some moles to track the truck routes and seem to have effectively disrupted the subterfuge. We'll see.

Figure 19.21. More activity at Bajo Flores.

19.24 SO MANY HOT PRESSES

September 27, 2007.

Whilst our partner Darko works away in Canada to try to get funding for future work (to build more hot presses in Africa), here in Buenos Aires, I try to understand his design and interpret it for the potential hot press manufacturers that I have been speaking to. Its not an easy task - I feel my materials' engineering background, together with my poor Spanish, leaves me disabled in the task of machine design. I have met some excellent mechanics and machinists over the last few days but as soon as I had the drawings in front of me, all the 'yes I can do that for sure' nods became, 'how does that work?,' 'I don't understand' and 'I can't do it unless I see the actual machine, the detailed instructions….' We hope, however, that we now have two potential builders for the machine and we press on with both. Today, I went to visit Victor who runs a huge workshop together with one partner and a handful of employees. He told me, he has been working as a metal worker for 14 years, and it quickly became evident how experienced he is. I was amazed by the size of the workshop and the extent of the equipment and capacity they have to create tools and machines of all kinds, but more so that they have built this from nothing in the last eight months. We spent a long time trying to understand the design, comparing it with the more usual four column varieties, which are extremely expensive.

Figure 19.22. Victor's warehouse.

We also compared it with the designs of an English model and a model developed by a former PhD student, Helen Cartledge and her brother, who has sent the designs to me by post from Australia. Victor was astonished at the amount of pressure we needed, and we had all sorts of fun deciding how the machine would actually work, but eventually, we got to a place where we think we can go ahead and start building. Tomas, our other hot press builder, is a designer who will work with machinists to realise the final product. He may come up with a different version. We have decided to go ahead with both - the race is on for the best hot press in town…In the meantime, I also visited INTI again today and finally met the Plastics group - and I saw my first working Buenos Aires' hot press. I am beginning to think that the only thing that has any real essence in this world is the hot press… Maybe I need a rest.

19.25 TRAVELING KNOWLEDGE

September 27, 2007.

We arrived in Buenos Aires almost 3 months ago (it's now the project's midpoint) with some fairly ambitious but, nevertheless, straightforward goals:

* Share knowledge of the processes of combining waste plastic with natural fibers to produce affordable composite building materials and domestic products.

* Use this knowledge to create opportunities for generating autonomous and sustainable revenue streams for some of Buenos Aires' most at-risk populations.

* Help reduce the damaging environmental impact of non-recycled plastic waste products.

* Create a 'how to' pictorial handbook (mass-produced and distributed royalty-free worldwide) that describes the methods, materials and manufacturing processes needed to combine locally collected waste with the fibers of naturally grown materials and transform them into useful building and domestic products.

We do have another, unstated goal, or perhaps it's just a question or a test or a self-indulgent exercise: Do two people, or three people, or whatever, have any chance of accomplishing 'development' (I use many quotes around this word) outcomes without the benefit of any pre-arranged alliances or financial or institutional support? We can't answer this yet, but we've certainly passed through some hopeful and less hopeful days.

Challenge Box: Well, do they?

If you've followed our blog posts, you probably have an idea why we chose Buenos Aires for this project: the political framework of the 2005 Zero Garbage Law; Argentina's long history of workers' collectives and more recent history of self-management, and a 6,000,000 people strong multi-reciprocal exchange economy that thrived for a few years; BA's large, informal, work force of cartoneros who earn very little money but have access to the plastic and cardboard needed to make the composite materials.

Before we left for Argentina, our close friend, Jackie, acted as an intermediary to give us the message, 'be careful, the Mafia controls garbage in Buenos Aires.' We have heard some anecdotal information that this is true, but it hasn't affected our work, though sometimes, we spend a little downtime speculating on what this sentence actually means. We have, though, been very quickly sucked into the politics and economics of waste, and it's awfully difficult to get a manageable picture of what's really going on here. This is what we think we know, though we're sure the reality is much messier.

*The Zero Garbage Law went into effect in late 2005 with the intention of 'reducing as much as possible the garbage that goes to the landfills or is incinerated, to curb pollution of the soil, air and water,' said Greenpeace activist, Juan Carlos Villalonga. The law stipulates that the amount of garbage in landfills is to be reduced by 50 percent by 2012 and 75 percent by 2017 from 2003 levels. One of its ancillary benefits is that recyclable materials will no longer end up in landfills.

* Garbage collection in Buenos Aires is almost entirely privatized: the city government contracts 5 different trucking companies, each of which is assigned a specific district, and they are responsible for cleaning the streets of waste and hauling that waste to transit points or directly to the CEAMSE landfill. The single government-owned 'company' hauls waste from Buenos Aires' poorest district in the southwest of the city.

* Buenos Aires does not yet have a visible and/or official systematic recycling program. We've recently seen brightly colored CEAMSE plastic and glass recycling receptacles in non-residential areas of the city and have heard that there is a plan to put one of them on every city block. Residents would be responsible for dumping their recyclables into the containers, which would then be picked up by CEAMSE and….brought where, sold to whom?

*Buenos Aires generates about 4500 tons of garbage everyday. 11% of that garbage is disposed of by the 4000-20,000 cartoneros (we've heard both figures and many others in between), the city's informal garbage pickers; 97% of the city's recycling is done by those same cartoneros who make 80-300 dollars/month by selling the recyclables per kilo to intermediaries who then sell the material onto recycling companies.

* There are several recycling cooperatives (usually made up of former cartoneros) which buy, process and sell some of the recyclables, though the economics of these cooperatives is very, very fragile.

*The municipal government does have plans to build 5 more sorting centers or 'green points' scattered around the city. One is currently active. Each center will be run by a different cooperative and receive waste hauled to it by one or more of the trucking companies. They can also receive recyclables brought to them by cartoneros in trucks (only). The sorted waste is sold somewhere up the recycling food chain. Bajo Flores is one of these centers, and is run by a 50-member cooperative. 6 centers will probably not be able to support more than 300-400 cartoneros by giving them the opportunity to turn their informal work into formal work.

* CEAMSE (Coordinacion Ecológica Area Metropolitana Sociedad del Estado) has been in business for 30 years and is the biggest player in this whole story. It is a municipal and regional government amalgam with private affiliations, which manages the landfills that receive the waste of the 13,000,000 people of the greater metropolitan area of Buenos Aires. No one we've spoken to has been able to unravel either the structure or operations of CEAMSE, and opinions vary widely

about whether or not it is the devil incarnate. CEAMSE has only one landfill that is in operation, Norte III, a little over an hour's drive from the city center. (The others are either being turned into 'eco-parks' or have collapsed.)

We are quite insignificant compared to the major players here, which works to our benefit, and sometimes we lose focus trying to figure out what the 'big' story is. But we have met some small collectives who are collecting and selling plastic or washing plastic or processing it, and who make a meagre living from waste but are eager to step up to small manufacturing. With the help of our young intern, Nils, who is beginning a cost-benefit analysis; and with Victor and Tomas who have taken on the challenge of building 2 hot presses; and the folks at INTI who will allow Caroline to experiment using some of their equipment with, hopefully, The Working World, who can help put this whole package together for us; and Carlos Levinton and his group who will keep this thing going when we've left, our pieces are coming together and our story is sounding quite good today.

19.26 COOPERATIVA DE TRABAJO "19 DE DICIEMBRE"

October 10, 2007.

Yesterday, accompanied by Rhiannon, our amazing friend and translator, we visited the Cooperativa de trabajo "19 de diciembre." We learned about this 'recuperated' metallurgic factory shortly after arriving in BA, and it's name kept popping up over the past few months. After years of layoffs and a final shutdown, the bankrupt factory (originally named ISACO) was occupied by a group of its workers on December 19th, 2002 (thus its name), who formed a cooperative to recover the company and get it working again. It continues, as in its heyday, to manufacture auto components for the automobile industry and, ironically, one of its principle clients is it's former owner who has created another factory which buys and assembles the components and sells them on to the large automobile manufacturers (Ford, VW, Mercedes Benz). The '19 de diciembre' employs about 5 or 6 outside skilled workers who work the more complex die machinery and are paid $4/hour, which is twice as much as the 30 or so cooperative members earn. (It is typical for skilled workers to either leave or not join cooperatives because they can make more money as 'independent' employees.)

We are trying to prepare for what happens when we leave Argentina, and one of our thoughts was to see if any of the recuperated metallurgic factories would be interested in manufacturing and marketing the hotpress, which is the key piece of equipment needed to make the fiber-reinforced plastic composites. Two hotpresses are in production right now and, if they work (fingers crossed), our group will have successfully designed a piece of $50,000 machinery for under $1000! We think it might make economic (and solidarity) sense for one of the recuperated factories to manufacture and sell a hotpress at a small profit to any of the recycling cooperatives who want to use one, and at much larger profit to the plastics industry where hotpresses are commonly in use. We spoke about all of this with Enrique Iriarte, the Cooperative's president. Now, it's wait and see if the idea flies.

19.27 INTI

October 13, 2007.

We recently developed links with Patricia Eisenberg's group at the Plastics Department of INTI (Instituto Nacional de Tecnologia Industrial). Initially, we hoped to be able to use the hotpress at INTI to try out some of the local materials, but today, we met Patricia, and she was keen to move beyond this and to work in a much more collaborative way with our project. She mentioned that there needs to be sustainability beyond our visit to Buenos Aires, something we have been losing sleep over, so it was like a dream come true to hear her words. We are very excited that the INTI group might be able to provide training and/or technical support to cooperatives and possibly research new waste materials which might be used and for which there is no current market. Patricia proposed a meeting in the near future to develop collaboration points between Waste-for-Life, the team at UBA and INTI. We've scheduled one for next Tuesday.

Another exciting part of today's visit was finally having access to a hotpress and trying out the methods developed by my Queen's student Ryan Marien, who has been volunteering on this project from afar. Using his method, Adrian, Mariana and I were able to re-create the cardboard fibre composite process with local materials – including the-standing-on-it-to-keep-it-flat-whilst-cooling part!

Challenge Box: What are your thoughts on sustainability of development projects beyond their expected duration - i.e., end of term, end of design project, end of PhD, end of funding?

19.28 SYNCHRONICITY

October 13, 2007.

It seems to me that the conversations we are having now with cooperatives about our project are more detailed and to the point than they were when we arrived. Of course, you will say, because you now know more about what you are doing. But in fact, I haven't noticed that we present the project in any different manner, but more so that the cooperative members ask much more sophisticated and relevant questions now, as if it's the fourth time we have been to see them, rather than the first. It's as if the idea is in the air…Some of the details that have been discussed recently are absolutely key to the success of the work within any of the groups we hope to work with. Our intern, Nils, is creating a cost analysis template and many of the factors we have been researching for this were brought up in recent meetings, such as 'How much electricity would the press use?' 'What could we sell the product for?' 'How long will a ceiling tile last?' 'Who would buy it?' These conversations took place in our meetings today and yesterday in Cooperativa El Alamo and Reciclando Sueños, both very successful and active cooperatives working in the recycling business. Both groups have an incredible spirit and believe strongly that they are providing a service just as the garbage trucks are, so they see no reason why the trucking companies are paid huge sums of money and the cooperatives are paid nothing or small subsidies. Marcelo from Reciclando Sueños had a head on fight with the government about this at the recent conference we attended on recycling.

Reciclando Sueños is the first co-op we have seen which has actually moved on to manufacturing from collecting, sorting and selling. They have an actual product – a painting sponge, the handle and backbone of which is made from recycled plastic. They chop, clean, dry and injection mould the plastic, and then assemble the parts, selling them for 1.10 pesos each to a local wholesaler. We were extremely impressed with their team, who were supported by weekly workshops run by local PhD anthropology students. After a long cold wait in a draughty warehouse full of bags of plastic, we were privileged to experience one of these workshops. It seemed to us to be a cross between a recovered workers assembly (although we have only heard about these) and a cooperative development workshop. Members were asked to discuss certain issues and topics, including a series of case studies, which led onto some incredible discussions such as 'How should we distribute the benefits gained when someone is given an extra item by the neighbours such as a good pair of socks?' Sebastian and Maria, who have been working with this cooperative for over three years, facilitated the discussions brilliantly.

Cooperativa El Alamo is to become one of the city's Green points. We were shown around the future premises, currently damp and mosquito-ridden, and proudly told which room would be the sorting and which the compressing area. We were introduced to both groups by Gonzalo Roque of Avina foundation, who took us to El Alamo today.

19.29 RECICLANDO SUEÑOS

Published by Rhiannon Edwards October 13th, 2007.

On a rainy Thursday, October 11, I accompanied Caroline and Eric to the cooperative Reciclando Sueños located in La Matanza, the largest zone of the Province of Buenos Aires with about two million inhabitants. The cooperative, which has collected recyclable waste in the nearby middle class neighbourhood of Aldo Bonce since 2004, stood out in a big way from others that we have had contact with.

The cooperative is the epitome of do-it-yourselfism. It was very evident that the members of the cooperative had an extensive knowledge of the types of materials that they were working with, but this is knowledge attained through trial and error and collectivizing the knowledge learned from successes and failures. They've managed to build a few major and complicated machines to process

- cut into small bits, wash, and dry - the recycled plastic that they collect. This is important because the more capacity the cooperatives have to process recycled materials, the less they are at the mercy of intermediaries and the more value they can extract from the material. Apart from the impressive machines made from scratch and salvage to cut, wash, and dry plastics, they've also developed an injector machine that melts and molds a certain kind of plastic and produces the base and handle for a painting tool, which they will be selling in the hardware stores of the neighborhood where they collect recyclables. This is an important development from many angles. Economically, it serves the cooperative because they can get a much better price selling a finished product to retail stores than selling the raw material to industry. They also spoke about producing these tools as being important symbolically in the development of their relationship with the residents of Aldo Bonce, because they can show not only that they are providing an environmental service by ensuring that recyclable materials do not end up in the landfill, but they are using the materials collected from the residents to produce something useful that returns to the residents. The injector machine is essentially the same idea as Caroline and Eric's hot press in terms of the function it serves for the cooperative; it's just that it produces a different tool.

This cooperative stands out especially for me because of its position in relation with the government (the municipal government is the relevant organ) and its vision of the work that they are doing. Reciclando Sueños defines their work as a public service that the government has failed to provide. They see no sense in the current arrangement of the government paying large companies to collect waste. The service they provide is one that serves the environment and the neighborhood, especially with the return of useful products made from what they collect. They have refused, up to now, to accept (or rather apply for) any form of government welfare checks or work plans, insisting that their work is legitimate and they should be paid by the municipal government for the services they provide. This distinction is important, and it seems to me, important that the money be paid to the cooperative where they can decide democratically how to manage it - maintaining control over decisions of investment, wages, etc., - instead of what seems to happen elsewhere, that the government gives money to individuals, funneled through the cooperative. Reciclando Sueños approaches their relations with the residents, from whom they collect materials, from along the same logic, seeing their work as a service, and residents' participation as that of a party receiving a service that duly corresponds to them, instead of as an act of charity.

We had the incredible and unique opportunity to sit in on their weekly "workshop" where they discussed and resolved issues of the cooperative with some facilitation by the anthropologists, Sebastian Carenzo and María Inés Fernández Álvarez. Far from being preachy or manipulative as I, ever the cynic might expect from middle class academics, it was an hour of mutual respect and cooperativism in practice. They kept their facilitation to a minimum, mostly directing any input towards ensuring that the more soft-spoken cooperative members were listened to and clarifying, mainly through repetition and little to no content shifting, proposals and arguments of the members.

19.30 PAINTING A PICTURE

October 22, 2007.

Every Sunday night in the Bronx, before going to sleep, I take a few bags of separated garbage that have been accumulating during the week and put them down on the sidewalk in front of my apartment - plastic, glass, cardboard. These bags are picked up sometime between midnight and 7 o'clock Monday morning by city garbage trucks and …. And what? Well, frankly, I have no clue what happens to my bags of garbage and, like most people, just assume that New York City somehow takes care of it all. My responsibility ends at the separation and putting out phase, which is about the only connection I have with the disposal of what I consume, a process that is for the most part completely invisible to me. Sometimes, I see people sifting through the garbage looking for 'returnables' that they pile into their own bags or hijacked shopping carts and take to the nearest supermarket to get 5 cents on the bottle. I have no idea what this exchange represents in terms of their monthly economy, but I can't help but recognize that some people are interrupting my 'orderly' chain of consumption and disposal, bringing it to the surface, and making a few pennies off of it. If you multiply this scenario thousands upon thousands of times, you get the picture of what recycling looks like in Buenos Aires; only in that city, it is rare than anyone separates anything …. except the cartoneros.

The 2001 Argentinean economic crisis created vast, overnight poverty throughout the country and an entire class of citizens who lived off of garbage, which they gathered, consumed and, when they could, sold. It is estimated that 100,000 people still live off of garbage in Buenos Aires Province, and 4000 - 20,000 of them come in to the city proper to pick through the 4500 tons of garbage that is left on its streets everyday. Prior to the passage of the 'Zero Garbage Law' in late 2005, it was public policy to haul ALL garbage to the several CEAMSE landfills located on the outskirts of the city. (Not surprisingly, these landfills were/are located in the poor or poorest areas surrounding BA, and it is their residents who suffer the polluted water, noise and air that are part and parcel of the impact these landfills have on the local environments.) Nothing was formerly recycled; everything was dumped into the landfills. Not much has changed.

The purpose of the 'Zero Garbage Law' is to reduce the total amount of garbage going into the landfills, and since almost half of that volume is composed of paper, cardboard, plastic and glass, it is these recyclables that are really the targets of the law. On the surface, this looks like a simple environmental issue that can be solved over time with some public education and a few public policies, but because so many people are actually already making a living, as meagre as that living may be, from these recyclables, this purely environmental issue is really all mixed up with a very visible social issue as well. 97% of all the recycling that is done in Buenos Aires is being done, informally, by the cartoneros, and one question that was vigorously pursued at the recent "1er. Foro y Congreso Internacional de Políticas de Reciclado en Grandes Urbes" was why the government was paying 5 private trucking companies to keep the streets clean and haul all of BA's undifferentiated garbage to the landfills and not paying the cartoneros who were separating and often processing the garbage for recycling.

The cartoneros or urban recoverers who fan out over the city everyday often arrive by train from the outskirts (this is important because these special trains are, as I write, being shut down, effectively, depriving many of the cartoneros access to the waste that is their livelihood). These are the people who actually manage BA's recycling and, as best as I can tell, they can be divided into three groups - gathers, processors, and gathers and processors. The groups can be highly organized cooperatives, or tiny family 'businesses' that include parents and their children, or people working on their own, and they may have ties with the neighborhood from where they collect the recyclables, or work anonymously and have none at all. Some of the garbage is sorted and sold 'as is' to intermediaries and some of it is processed - either washed, or squashed or sliced and diced, and sold a little further up the food chain. But we're talking about pittances here, and at an average of about 500 pesos per month, most cartoneros are part of the 2.4 million people or 20% of GBA (Greater Buenos Aires) who live below the poverty line, which the government has pegged at 914 pesos per month. Our tiny intervention here could add a small, local manufacturing stage to the recycling process, which, if managed properly, would supply another source of income for the cooperatives.

19.31 PRESSING AHEAD

October 26, 2007.

There have been more than a few unsettling moments during the past 3 1/2 months. Most of them have to do with the recognition that we have round-trip tickets and will be leaving here, that we will be alright. But the people we work with and have grown so close to can't really go anywhere, and their futures are much, much less certain. Perhaps, the biggest drag on moving forward, the thing that has the potential to finally torpedo everything we've been doing here and turn Waste-for-Life into an academic exercise, has been the hotpress - or lack of a hotpress - which is the key mechanical technology we're sharing with the recycling cooperatives. Getting the hotpress manufactured here (as well as in Canada) has been a major stumbling block, but yesterday, we visited Tomas' workshop, which was a 3 hour trip for us out of the city, and saw the pieces of the hotpress ready to be assembled. We expect that we can actually begin training sessions within the month.

19.32 THE PHOENIX OF LANZONE

October 26, 2007.

Renacer Lanzone or the Pheonix of the barrio Lanzone is a social organisation run by Adam Guevara. Adam had an idea over twenty years ago, that CEAMSE should send the trucks to groups of cartoneros, before putting the garbage in the ground, so that they could take the recycling materials, and leave less waste to be buried. He tried with many different Presidents of CEAMSE but it was not until the present one agreed, over three years ago, that Adam was able to realise his dream and set up a social factory or sorting centre in Lanzone district. His project was so successful there are now four centres and four more planned on the CEAMSE site – some of which we had already seen. Adam's group are very successful, he told us, because they have been trained by INTI - the

Figure 19.23. Buenos Aires hotpress in the making.

same plastics group that we work with, to separate the plastic waste very well and to sell each type individually. They make a lot more money than the other groups as a result – up to 20 pesos an hour. Adam cares very much for his group, and when I asked what criteria he had for choosing people to work there, he told me it was on the basis of need – if they have more children, etc. He even keeps a section of the factory as a store of PET bottles to become a Christmas bonus at the end of the year.

Training seems to be an important part of the success here. INTI training the technical selection process and also Suarez, a local professor, and Neumann, a consultant engineer, working for the University of Sarmiento and CEAMSE, training them in roles and responsibilities, team work and conflict resolution, the context, history of recycling and cartoneros as well as the more technical matters. Adam is very keen to move on with us, and he seems to be an extremely innovative thinker. As wary as we are of CEAMSE, we would love to work with Adam's Pheonix.

Challenge Box: At this point, with six weeks to go, we had no idea whether the project would work, whether we would have to come back, whether INTI would support the work, if the hotpress would work, if the cartoneros would find a product to manufacture.... What would you do next?

19.33 THE MIGHTY HOTPRESS IS FINISHED (ALMOST)

November 17th, 2007.

We passed by Tomas Benasso's workshop a few days ago to gaze at the finished hotpress and go over some of the minor kinks that he's going to work out during the next 2 weeks. It's a mighty beast of a machine and a tangible measure of one of the things we're trying to accomplish here in BA. The very good news (besides Tomas' heroic work from Darko's design) is that INTI's Plastics Group has agreed to house the Kingston Hotpress and begin putting it through it's paces, which will include property analysis and testing samples against local building and product codes and standards.

Figure 19.24. The hotpress in Tomas' workshop.

19.34 SIN PATRÓN

December 12th, 2007.

While waiting for the final tweaks to the hotpress (by the way, Tomas has tweaked it and is delivering it to INTI this Thursday, December 13), the remarkable Erika Loritz has helped us conduct street interviews with cartoneros.

We have spent most of our time in BA working with cartonero recycling cooperatives. They have the structures in place to successfully incorporate a small manufacturing channel to their, mostly collection, sorting, processing and distribution operations. But equally important to us is a common political and social outlook that stresses equity and human interdependence, and we have been the privileged witnesses of these ideas in motion, particularly their commitment to wage equality, non-hierarchical participatory decision making practices, and to 'socializing' their knowledge. We did not come to Buenos Aires to be community organizers, and so the individual or small family groups - the informal cartoneros who do much of BA's recycling - have been, for the most part, a closed book to us. We realized, early on, that we were not hearing their voices, although we did hear many opinions about them from other people, but we didn't have either the language skills or the confidence to simply go up to these people on the street and start talking. I spoke about this with Erika in Iruya, and she immediately offered to help us take up this challenge.

We have done 4 interviews so far and have others to do, so we are not working with a large dataset; however, there is a strong leitmotiv that we can point to, that we've been aware of ever since attending The Workers' Economy: Self Management and the Distribution of Wealth conference in July, and that can be summed up with the phrase, 'sin patrón.' This phrase, which means 'without a boss' or 'without an overseer,' gained currency during the recovered factory movement that began with the 2001 economic crisis, but we heard it used over and over again by the cartoneros we interviewed to describe the essential quality of their lives in very positive terms. They did not want to work for anyone, they did not want to be subject to the collective decision-making processes of a collective, they did not want to bother anyone, and they did not want to be bothered. It is still much too early to sum up the content of the interviews, and we have a lot of transcribing and thinking to do, and loads to read up on about informal work, but with Erika's help we have a special opportunity to learn something about this previously unavailable population of urban recyclers.

19.35 CHAMPAGNE!

December 16th, 2007.

Tomas delivered the Kingston Hotpress to INTI on Thursday as scheduled, and after about 2 hours of fiddling around with the electrical installation, we had a go of it with the strips of plastic bag that Caroline had been cutting up all morning - a task that was as much a necessity as occupational therapy, seeing that we were all sitting around on the edges of our seats. The hotpress had never been road tested - it only worked, theoretically, according to Darko's calculations and ingenious SolidWorks design. This was a lean and mean machine and, in order to keep the costs down, had no gauges (gadgets) to let us know that it actually generated enough heat (160°C) and enough pressure to melt the plastic. From my non-scientific mind's perspective (after all, this was only the second time I had ever been in a research laboratory) the first test was a disaster. We were all much too impatient and didn't let the press heat up enough - we expected this would take about 20 minutes, but only gave it 10 - and didn't leave the plastic in the press long enough - we thought this would take about 15 minutes, but only gave it 7. We cooled the mold down with water and opened it to a soppy, un-melted plastic mess. Our faces were all frowns. At this point, Patricia Eisenberg leaned over to us and said, 'don't worry, we'll make it work,' and her INTI crew stepped in, inserted a funny little L-shaped thermometer into a gap in the press, declared that it was indeed hot enough to melt plastic, poured a bucket of polyethylene pellets over the few strips of plastic bag that remained in

Figure 19.25. The hotpress in the lab at INTI.

the mold, spread the pellets out evenly (kind of) over the 60 cm × 60 cm surface, and shoved the mold back into the press. Tomas pumped up the car jack that created the pressure, and I was the timekeeper. 15 minutes later, Tomas released the jack's pressure, and we took the mold out of the press, laid it on the sidewalk just outside the building's side door, hosed it down, and opened the mold to reveal a slightly rough around the edges but beautifully hardened 2 kilo plastic plaque. The hotpress worked, just as Caroline had always believed, said, knew it would, and we've invited members from the 9 cooperatives we've worked with during the past 6 months to come by INTI on Wednesday to see it in action. Six months after arriving in BA, it's time for champagne!

19.36 HANDING IT OVER

December 19th, 2007.

    In a few hours, we're off to INTI to participate in what is surely our final Waste-for-Life meeting before leaving Buenos Aires. We've already begun saying goodbye to our many 'compañeros,' a word that after 6 months of learning and struggling here has real meaning for us and is not at all embarrassing to use. And though this may be our last meeting, it is really a first because we are bringing together in one place all of the players who will keep Waste-for-Life BA alive after we return to North America - representatives of the 9 cooperatives we've worked with since July 2007; Carlos Levinton from the University of Buenos Aires' CEP; Gonzalo Roque from the Swiss NGO Avina, and Patricia Eisenberg's INTI plastics group. (The only person who can't make it today is

Figure 19.26. The first misshapen attempt at molding.

Esteban Magnani from the micro-credit organisation The Working World (La Base), who we met with yesterday and proposed a collaboration scenario which he enthusiastically endorsed.)

We've just returned from the meeting which, because of the collective engagement and emotion, caught us pretty much off guard. We had clearly handed over Waste-for-Life to the local stakeholders and could feel OK about going home. There was a single moment in that research laboratory, amidst the hotpress and all of the other grey testing equipment, and all those people who had such different stories to tell, but each of whom had been drastically affected in one way or another by the last decade of Argentinean history, when it became crystal clear why we were doing what we were doing. Adam Guevara talked about what he had learned from the INTI scientists and how it had changed the life, yes, the life, of the 20+ members of Renacer Lanzone, the civic association that he runs. These people collect and separate and sort plastic, which is a stinking job, and a couple of INTI scientists from Patricia's plastics group had spent time with them, some time ago, teaching the group how to do their job better by being more precise in the identification and, thus, the separation and sorting processes. Adam's group, whose members come from the shantytowns across the highway from the huge CEAMSE dump, took these lessons seriously, and now are able to sell their plastic for 4 times as much money as any other recycling group we've met. It was Adam's first opportunity to thank the INTI scientists, which he did with great dignity, and which sent some of them out of the room in embarrassed, unexpected tears.

Figure 19.27. The team at INTI - looking at the press.

19.37 WHO'S DOING WHAT?

December 19th, 2007.

Who's doing what? How will Waste-for-Life BA sustain itself? The only thing we knew for certain before coming to Argentina was that our stay was finite. The time that separated our flight into and out of BA was 6 months; we knew very little else. It's premature to map out what we've learned and done, and way too early to draw any conclusions, but one thing we can point to is that we've adhered to our idea of Waste-for-Life as a loosely joined network of people with diverse competencies, sharing common values, who join together to work through poverty-reducing solutions to particular ecological problems. We had no a prior idea what form this would take, but this is what the decentralized structure looks like to us as we prepare to leave.

Levinton's CEP is taking care of the experimentation in BsAs. They will spend the next few months putting the hotpress through its paces using different types of plastic waste in combination with cardboard and other natural fibers that they will collect from the cartonero cooperatives. They will test the results for compliance with local building codes and standards and will work alongside the cooperatives to teach them how to make the reinforced plastic composites. The project will be supported by our research team in Canada and once we have nailed down the technical processes, CEP will begin product development with their own hotpress, seeing what useful and/or fanciful products they can tease out of it. They will work closely with the cooperatives, receiving and testing out their ideas and helping their members with production. Avina is waiting in the shadows for one of the cooperatives they work with to propose a manufacturing and distribution plan that they can support with their funding and facilitation resources, and it is very likely that they will be behind the first complete test case. The Working World (La Base) will help identify and work with the cooperatives, which have the most likely chance of developing a successful manufacturing channel, and will establish and administer a revolving hotpress loan fund to enable the cooperatives to purchase a hotpress. And the cooperatives will be the innovators, the inspired ones, for they are the ones who have hope, who struggle every day, who work so hard with so very little yet can stand on their own shoulders and see beyond what they do now into a future where they deserve and will have a little bit more.

19.38 FINAL THOUGHTS

This has been rather an unusual chapter. It comes, as you can tell, directly from a blog so you have been able to track the movements of our team as we worked with the cartoneros in Buenos Aires. It is clearly an ongoing project. Many questions have been raised and many left unanswered for you to ponder. There are no right answers about any of these issues – but there are ways of thinking that are more helpful than others. Making sure that you think things through and question your prejudices and assumptions will always be a good thing to do. We had to do this constantly. Even as we left BA and sat in the airport wondering at what we had done, we had many elated thoughts such as 'How did we ever get to elicit the help of such amazing people?' 'How does Adam Guevara develop such a giving attitude when he has so little?...' We also had some troubling thoughts – 'Would the groups, together with the support of the University and INTI, actually be able to manufacture a product worth selling? Would they be able to sell it? Would we be able to get an organisation to make more presses? Would this make a difference to the cartoneros' lives as the knowledge about sorting plastics, that Adam's group had gained from INTI a few years before had done? Were we leaving too soon or was it time to go and hand over the tiller? None of this seemed to matter. We had handed the project over without even knowing it. We were never in control of this work. There were so many factors, so many features to think about, and we had tried to react to everything and think of as much as we could. We were delighted when Estaban from Working World said to us a few days before we left – 'you did it you know, many come here with great ideas and they just don't work.' But what had we done and how had we done it? We asked ourselves this question time and time again. The lessons learnt were many, but in the end, we decided the most important one of all was that we had listened; we had sought out as many different voices as we could and we tried to remain unbiased whilst staying within our non negotiables. These were: working with a participatory, inclusive approach, with the aim of poverty reduction, and the enhancement of social justice and environmental sustainability. We felt that all the various players who had learnt over time to trust us and to work with us had understood that this and nothing else, other than our own personal satisfaction with doing what we enjoyed, was all we had to gain from this project.

> **Challenge Box:** Think about what the team achieved in the six months. What more could they have done? What could they have done differently? What should they do now to make sure that the next steps work?

C H A P T E R   20

Turning on the Tap

Figure 20.1. Funnelling public water for private sale.

20.1 TURNING ON THE TAP

Its very easy to imagine, when we are living in England, the U.S., Canada or other parts of the world where water is in abundance, that everyone can just turn on the tap as we can and water will flow. However, we are starting to understand that water is not always available and that we need to be careful with how we use it. Australia has seen a drought for several years at the time of this writing and Australians are now shocked to see water being used to water gardens in other countries as this was banned a long time ago in their country. In England, a country, which seems to see rain all the time (although I suppose it actually only dribbles), we had a drought in the 70's which meant that we had to use pipes in the road as water was cut off to individual houses. In Book 1, Chapter 4, we saw what happens when droughts are badly managed and the devastating effects of famines come upon countries. 'Water scarcity may be the most underappreciated global environmental challenge of our time' (Barlow, 2003, p3), (World Watch Institute, 2006).

Figure 20.2. Children playing in fountains in Salt Lake City.

Young people who have not directly encountered droughts of any degree can easily ignore the sustainability of water as something not relevant to them. I once asked a class of students to write about creating a sustainable city - what would they need to consider and how would they do this. Despite the clue of the project topic, all 600 students, when I asked them to describe 'water-use' in their city, described how they would create marinas and water sports facilities. When I spoke to one of the university governors about this in passing, she said 'Oh yes, my children would be the same as they are so protected and lucky here in Canada.' The very important topic of sustainability is not one we will focus on in the section, although it will arise in the discussions. We are going to look in more detail at the issue of privatizing water. When I was young in England, we did not have to pay for water. Water was a free resource. It still is to me. I was shocked when we had to start paying for the amount if water we used. It just felt wrong. In the discussion below, we look in more detail at the differing views and implications of privatizing water in different parts of the world.

We will see that there are many different views on this topic, and water privatization has had a powerful effect on the way we think about ownership of natural resources. Some say we mine oil

Figure 20.3. Blue gold.

and gold and we sell that and we also sell trees, which grow on our land so why not sell water too. Others, the authors of this book included, feel that it should be a basic human right. Suffice to say that there are many people in the world without access to water, either because their village has not yet been developed enough to be connected by pipes and pumps to a water source, or because they cannot afford the pipes, or because they cannot afford to turn on the tap, and it has been cut off by the company who now 'owns' the water in their area.

> **Challenge Box:** Track your water consumption for one day and compare with your friends. If possible try to find out how this compares with someone from a country where water is less abundant. What differences are there?

20.2   BLUE GOLD

In Canada, Maude Barlow is something of a heroine, fighting for the rights to water. She has written some very useful texts, 'Blue Gold: The Fight to Stop the Corporate Theft of the World's Water' and 'Blue Covenant: The Global Water Crisis and the Coming Battle for the Right to Water' (Barlow, 2003, 2008). Her basic facts are helpful and continue to amaze us. 'The average human,' she tells us, 'needs fifty litres of water per day for drinking, cooking and sanitation. The average North American uses almost six hundred litres a day. The average inhabitant of Africa uses six litres per day' (Barlow, 2003, p5). But how can we help? Think most Canadians. We hear: 'Its not my fault that we have more water here.' But virtual water is a huge problem. This is water, which is displaced through the trade of goods which needs water to grow or manufacture them. Marlow tells us that Kenya is using up the water of Lake Naivasha to grow roses for Europe. Biofuels, for all their environmental promise, use up land which used to be used for crops, but also use massive amounts of water. It takes 1,700 litres of water to produce 1 litre of ethanol. 'Vietnam is destroying its water table to grow coffee for export' (Barlow, 2003, p17).

But what has this got to do with engineering? Sadly, everything – in many direct and indirect ways. The technology used to manufacture the biofuels was developed by well meaning scientists and engineers thinking that they were helping to reduce global warming but without fully understanding the implications of their creation. The huge increase in manufactured goods uses up water, which we don't even think about. Bottled water is one of the most polluting industries that exist. They have become a fashion accessory and healthy outdoor types love to carry a bottle around but the reality is that one million bottles of water cause emissions of 18,000 kg of carbon dioxide and fewer than 5% of bottles are recycled (Barlow, 2003, p99). I trained in materials' engineering, but I did not learn how much water was needed to create the plastics and metals, which we make everything from.

Even when strategies have been developed to try to alleviate the water crisis, Barlow informs us that the high technology solutions, the dams, diversions and desalination are part of the problem. Larger dams 'significantly contribute to the emission of greenhouse gases, and therefore to global warming, one of the greatest threats to freshwater resources' (Barlow, 2003, p22). Diversions where water is carried in giant pipes, Barlow informs us, will result in water depletion in the long term. Many poor rural areas become deserts when water is 'sold, expropriated or just plain stolen' (Barlow, 2003, p23). The final technology being used is very expensive and very energy intensive – desalination. It creates a lethal by-product and other contaminants which are pumped back into the sea. Furthermore, sewage is often dumped into the sea, which is then drawn back into the desalination plant for human consumption. The very new technologies such as Atmospheric Water Generators and Cloud Seeding are privately owned, allowing the companies to develop the market without competition.

Clearly, with the inevitable water crisis looming, water becomes a commodity worthy of value. However, the shift from a public to a private model is traced back, not too far, to Margaret Thatcher's Britain, and Ronald Reagan in the US. British publicly owned water was sold off to private companies at bargain prices. Millions had their water cut off. With the increasing neoliberal policy adopted by the World Bank of forcing developing countries closer to a market driven economy, water services in the Global South became targets for privatization.

Barlow explains how the World Bank and big water companies set out to promote a major shift in water policy, seeking buy-in of NGOs, think tanks, media and the private sector. 'Through its Water Capacity Building Program, the World Bank Institute (the capacity building arm of the bank that promotes bank values and programs through education and outreach) has put thousands through intensive programs on private water management' (Barlow, 2003, p42).

'For with the acceptance of water as a commodity comes the dilemma of what to do with the idea of water as a basic human right. In other words, if we are willing to use monetary value as our sole guiding principle for water extraction, treatment and distribution, on what grounds do we make moral decisions about how much water is enough and who is consuming too much? Just because someone can afford to pay the cost of filling their swimming pool or washing their cars every day should they have the right to do so when others are struggling to survive with no water at all? (McDonald in Francis, 2005, p 33)

Francis (2005) tells us that in 2001, South Africa adopted a policy of Free Basic Water (FBW), which aims to provide each household with 6000 liters of clean water every month free of charge.

This is deemed the minimum amount needed for survival. One of the strongest critics of FBW is that the quantity defined by the government is insufficient - 33 l per day compared with the UN suggested 50 l for a household. Also because it is per household – poorer families tend to have larger numbers in a household. Since 2002, there is a justiciable right to water. But what does this mean? It places an obligation on the government to take reasonable action to achieve these rights. The problem is that there needs to be a commitment of financial resources and cooperation amongst federal and local officials to achieve anything useful with this, and to date, this appears not to be the case.

Challenge Box: Do you think water should be free for all? Justify your response.

20.3   THE WATER WARS

The term 'water wars' referred originally to the famous protests in Cochabamba, Bolivia, in the year 2000, when the government sold the municipal water supply of the city to an international water company. There were many street battles and loss of life with huge debates worldwide. But what are the underlying reasons for disagreements on water, amongst those who proclaim it to be for poverty reduction? Kanbur's understanding (Kanbur, 2007) of why people might be against water privatization is that the selling of some things such as slaves and child labour is intolerable to them, and this must also be the case with water. However, there is a much deeper political divide in most of the papers you can find on the subject. Either you believe that the government should look after our basic services, or you believe that private industry should do this. In some cases, we find alternatives to this but not many.

Kanpur picks out a debate between Vandana Shiva and Fredrik Segerfeldt, which he quotes from extensively, and we will do so again as it illustrates the differences very well.

'The World bank started to push the Delhi government to privatize Delhi's water supply…The contract between Delhi Jal Board… and the French company Ondeo Degrement (subsidiary of Suex Lyonnaise des Eaux Water Division – the water giant of the world), is supposed to provide safe drinking water for the city…On December 1, 2004, water tariffs were increased in Delhi. While the government stated that this was necessary for recovering costs of operation and maintenance, the tariff increase is more than ten times what is needed to run Delhi's water supply. The increase is to lay ground for the privatization of Delhis water and ensure profits for the private operators…the tariff increase is not a democratic decision nor a need based decision. It has been imposed by the World Bank. ….How will they give water to the thirsty? Cremation grounds, temples, homes for the disabled, orphanages which paid Rs 30 will now pay thousands of rupees. …The World Bank driven policies explicitly state that there needs to be shift from the social perception to a commercial operation. This worldview conflict lies at the root of conflicts between water privatization and water democracy. Will water be treated as a commodity or will it be viewed as the very basis of life? The common argument for privatization and price increase is that higher costs will reduce water use. However, given extreme income inequities a tariff increase that can destroy a slum dweller or poor farmer is an insignificant expenditure for the rich. Privatisation as dictated by the Asian Development Bank / World Bank thus means that water will be diverted from the poor to the rich, from rural to industrial areas… Water privatization aggravates the water crisis because it rewards the waste of the affluent, not the conservation of resource of prudent communities. Sustainable and equitable use needs water democracy not water privatization.' Shiva, quoted in Kanbur, 2007

Figure 20.4. Vandana Shiva.

Ninety-seven percent of all water distribution in poor countries is managed by the public sector, which is largely responsible for more than a billion people being without water. Some governments of impoverished nations have turned to business for help, usually with good results. In poor countries with private investments in the water sector, more people have access to water than in those without such investments. Moreover there are many examples of local businesses improving water distribution. Superior competence, better incentives and better access to capital for investment have allowed private distributors to enhance both the quality of the water and the scope of its distribution. Millions of people who lacked water mains within reach are now getting clean and safe water delivered within a conventional distance…The main argument of the anti-privatisation movement is that privatization increases prices, making water unaffordable for millions of poor people. In some cases it does, in others not. But the price of water for those already connected to a mains network should not be the immediate concern. Instead we should focus on those who lack access to mains water, usually the poorest in poor countries…They usually purchase water from small time vendors, paying an average of 12 times more than for water from regular mains and often more than that. When the price of water for those already connected goes up, the distributor gets both the resources to enlarge the network and the incentives to reach as many customers as possible. When prices are too low to cover the costs of paying new pipes, each new customer entails a loss rather than a profit, which makes the distributor unwilling to extend the network. Therefore, even a doubling of the price of mains water could actually give poor people access to cheaper water than before. True many privatizations have been troublesome. Proper supervision has been missing. Regulatory bodies charged with enforcing contracts have been non-existent, incompetent or too weak. Contracts have been badly designed and bidding processes sloppy. But these mistakes do not make strong arguments against privatization as such but against bad privatizations. Let us therefore have a discussion on how to make them better instead of rejecting them altogether. Greater scope for businesses and the market has already saved many lives in Chile and Argentina, in Cambodia and the Phillipines, in Guniea and Gabon. There are millions more to be saved. (Segerfeldt, quoted in Kanbur, 2007)

Figure 20.5. Fredrik Segerfeldt.

Barlow does not need a debate. She is convinced. 'Almost twenty years of documented cases of the failure of privatization and growing opposition to the World Bank and the water service companies in every corner of the globe have revealed a legacy of corruption, sky- high water rates, cutoffs of water to millions, reduced water quality, nepotism, pollution, worker lay-offs and broken promises' (Barlow, 2003, p58). But the main issue that Barlow has is that water companies are for profit 'the ultimate goal of private companies is to make profit, not to fulfill socially responsible objectives such as universal access to water. In countries where most of the population earns less than two dollars a day, notes Sara Grusky of Water Watch, private companies cannot meet shareholder obligations to provide a market rate of return. Nor can they expand their services to a population that cannot pay. The only way that the private sector can stay competitive in such a situation is to have access to public subsidies, the very thing they were brought in to relieve' (Barlow, 2003, p58).

20.4   WATER IN BUENOS AIRES

Figure 20.6. Shantytown in Buenos Aires. Many residents of the shantytowns do not have running water.

Buenos Aires is an often quoted example of the water wars debates. It is one of the largest concessions to date - the major shareholder being the French company Suez Lyonnaise des Eaux mentioned by Shiva above. Privatisations can be management contracts, lease contracts or full concessions to the private company. In our discussion below, the first case comes from a group of authors who are all associated with the World Bank and present the case for privatisation in Buenos Aires. The second group represents a group of academics in the north and south who work on what they call the 'Municipal Services Project.'

Abdala (1997) suggests that prior to reform, water and sewerage services provided by Obras Sanitarias de la Nacion in Buenos Aires suffered from a lack of investment and inadequate maintenance.. 'overstaffing, unresponsive customer service, high levels of unaccounted for water (UFW) and low collection rates. Privatisation on the other hand brought about positive changes, resulting in rapid improvements in the performance of the company now called Aguas Argentinas' (Abdala, 1997, p4).

He concludes that the government came out as a loser but that employees in turn became beneficiaries as they received ownership of 10% of AA shares. Abdala states 'the result applies to those who remained employed and dividends could not be expected sooner than 1999. For this, workers who entered into the voluntary retirement programmes we assume that the severance payments that they received were enough to compensate for their disutility from being laid off. Buyers also 'came out with welfare gains' and consumers 'in turn are the big winners.' Existing customers received the largest share of this pie as there were important cuts in the average price for usage. New users also benefited as privatization reduced rationing by increasing access. Consumers also received benefit through improved quality effects and public health that we were unable to quantify. On the quality side, this is clearly the case of improved water pressure levels waiting time for repairs and customer service' (Abdala, 1997, p25).

Furthermore, Galiani et al. (2005) found that child mortality fell 5-7 % in areas that privatized their water services overall and that the effect was largest in the poorest areas. 'In fact, we estimate that the privatization of water services prevented approximately 375 deaths of young children per year' (Galiani et al., 2005, p1). They state in their conclusions that their results shed light on two important debates. The first they describe as follows 'One concern that has postponed privatization of water systems around the world is the fear that private operators would fail to take into account the significant health externalities that are present in this industry and, therefore, under- invest and supply sub-optimal service quality. Contrary to this concern, we find that the effect of privatization on health outcomes has been positive. Private operators have accomplished the network expansion and quality requirements specified in the privatization contracts or at least their level of attainment has been superior to the performance under public management. While the private sector may be providing suboptimal services they are doing a much better job than either the public sector or the non profit cooperative sector' (Galiani et al., 2005, p28).

The second area is described equally strongly. 'There is a growing public perception that privatization hurts the poor. This perception is driven by the belief that privatized countries raise prices, enforce service payment and invest only in lucrative high income areas. Instead, we find the poorest populations experienced the largest gains from privatisation in terms of reduction in child mortality. Privatisation appears to have had a progressive effect reducing health inequality' (Galiani et al., 2005, p28).

Some groups acknowledge that there were problems but do not believe that privatisation itself is to blame. The signing of a concession contract for the Buenos Aires water and sanitation system in December 1992, attracted worldwide attention, and caused considerable controversy in Argentina. The concession was implemented rapidly and, according to Alcazar et al. (2000), reform generated major improvements in the sector, including wider coverage, better service, more efficient company operations, and reduced waste. 'Moreover, the winning bid brought an immediate 26.9 percent reduction in water system tariffs. Consumers benefited from the system's expansion and from the immediate drop in real prices, which was only partly reversed by subsequent changes in tariffs, and access charges. And these improvements would probably not have occurred under public administration of the system.' Still, the authors show that 'information asymmetries, perverse incentives, and weak regulatory institution' could threaten the concession's sustainability. 'Opportunities for the company to act opportunistically - and the regulator, arbitrarily - exist, because of politicized regulation, a poor information base, serious flaws in the concession contract, a lumpy and ad hoc tariff system, and a general lack of transparency in the regulatory process. Because of these circumstances, public confidence in the process has eroded. The Buenos Aires concession shows how important transparent, rule-based decision-making is to maintain public trust in regulated infrastructure' (Alcazar et al., 2000).

McDonald and his group, on the other hand, take the opposite perspective (Loftus and Mcdonald, 2001). They explain that President Menem rushed through the national Administrative Reform Law declaring a state of economic emergency with regard to the provision of public services. The law authorized the 'partial or total privatization or liquidation of companies, corporations, establishments or productive properties totally or partially owned by the state, including as a prior requirement that they should have been declared subject to privatization by the Executive Branch, approval for which should in all cases be provided by a Congressional law. Through such a decree, Menem was able to privatize the Buenos Aires water and sewerage network Obras Sanitaras de la Nacion (OSN) without pubic consultation, arguing that it was 'urgent' to press on with reforms' (Loftus and Mcdonald, 2001, p11). Even amongst union leaders, there was a feeling of inevitability and that fighting was futile. The hyper-inflation which happened during this time in Argentina 'essentially disciplined the population into accepting privatization as a solution' (Loftus and Mcdonald, 2001, p12).

Table 20.1 shows the coverage at the start of the concession in 1993 [reproduced from (Loftus and Mcdonald, 2001, p14)].

Of the 30% population not connected to the water network in 1993, 95% obtained water from wells with pumps, and for those with sewerage services, 88% disposed of this through cesspools and the rest directly into rivers.

Table 20.2 shows the five yearly performance targets [reproduced from (Loftus and Mcdonald, 2001, p18)].

By 1999, Aguas Argentinas claims that water coverage had reached 82.4% and this meets the performance targets. However, they were at 61% for sewerage. Sewerage infrastructure had not kept pace with water delivery expansion and it is claimed that this is because water delivery is twice as profitable as sewerage for Aguas Argentinas. 95% is still dumped into the Rio Del Plata. We have

Table 20.1: Water coverage before the concession in Buenos Aires (Loftus and Mcdonald, 2001)

Water System	
Number of Connections	1,170,000
Average Production (m^3/day)	108,950,000
Treatment Capacity (m^3/day)	3,640,000
Length of Water Pipe System (kms)	11,000
Total Population	8,580,000
Served Population	6,000,000
Sewerage System	
Number of Connections	700,000
Served Population	4,700,000
Volume Collected (m^3/month)	82,232,000
Volume Treated (m^3/month)	3,413,000

Table 20.2: Five yearly performance targets (Loftus and Mcdonald, 2001)

Year of conces.	% population served by water	% population served by sewerage	% Collected sewage treated		Network renovation cumulative %		% unacc. for water
			Primary	Secondary treatment	water	sewage	
0	70	58	4	4	0	0	45
5	81	64	64	7	9	2	37
10	90	73	73	14	12	3	34
20	97	82	82	88	28	4	28
30	100	90	90	93	45	5	25

been told by the locals that the problems with privatisation follow from the fact that the higher water tables are now completely polluted as a result of the untreated sewage.

Challenge Box: Try to find out when and if water in your city/country has been privatized. What are some of the popular believes about the costs and benefits of this?

Figure 20.7. Rubbish tips and water ways in Buenos Aires. (The bicycles line up as cartoneros (see the chapter 'Throwing Away My Rubbish') wait until they are allowed into the landfill site to collect anything they might be able to sell.)

Apart from service, the main arguments for privatization are cost and accountability. It is usually said that private companies are more efficient than the public sector and reduce costs to the end user. Here, Loftus and Macdonald believe claims have been exaggerated. Costs went down 27% and then went up 20% - critics apparently argue that the higher prices were made just before so as to make the company look good. 'The effect of the increases (prior to privatization) was to allow the company to offer what seemed to the public to be a 27% decrease in costs, even though in reality it was a manufactured reduction' (Loftus and Mcdonald, 2001, p19).

It is also said that privatisation generates better public accountability. In fact, Loftus and Macdonald suggest that the opposite has happened. 'Starting with a Presidential decree in 1989 which unilaterally declared that the city's water and sanitation would be run by the private sector, all decisions about the extent and scope of privatization in these sectors were made behind closed doors. There was no public debate on the matter, and the first (and only) public consultation did not take place until June 2000, seven years after the concession had begun' (Loftus and Mcdonald, 2001, p3).

Loftus and McDonald summarise by stating that 'although there have been some impressive gains in the extension of water infrastructure, the majority of the concessions' negative impacts have been mostly felt by the poorest sections of BA. Many poor households have fallen into serious arrears and have been disconnected from the network, especially prior to 1998. …environmentally, those living in the poorest areas of BA have also been faced with the negative effects of rising groundwater and the health risks associated with nitrate contaminated aquifers. These municipalities have some of the lowest average incomes in the Greater BA and yet a large part of the financial burden for extending the network has fallen on these households' (Loftus and Mcdonald, 2001, p29).

As Kanbur (2007) puts it, 'Two more opposing descriptions of outcomes could hardly be possible. According to one analysis, the privatization was an unmitigated disaster. According to another, it saved children's lives' (Kanbur, 2007, pg8). These views, he says, are taken by two general tendencies which he calls the Civil Society tendency (CS) and the Finance Ministry tendency (FM). Kanbur gives away his preference by suggesting that the FM framework is more generally open and more humble about their findings and frameworks (Kanbur, 2007, pg11); however, his key interest is to 'understand the different methods of analysis and different ways of formulating questions. After a few such meetings, we would be in a position to judge whether the divide can be bridged' (Kanbur, 2007, pg11).

On March 21st 2006, the Argentina Government rescinded the thirty year contract of Argus Argentinas. An April 2007 report by the city's ombudsman stated that most of the population of 150,000 in the southern district of the city lived with open-air sewers and contaminated drinking water (Barlow, 2007, p106). Suez was also forced to abandon its last stronghold in Argentina, the city of Cordoba, when water rates were raised 500% on one bill' (Barlow, 2007, p107).

20.5 WATER IN AFRICA

We see a similar situation in studies of water privatization in Africa although it has been slower to develop. By the end of 2000, 93 countries had privatized some of their water services but only 7 projects in the Middle East and North Africa and 14 in sub-Saharan Africa. Kirkpatrick et al. (2004) suggest that across Africa there is better performance in private utilities compared to state owned utilities, but there are found to be no statistically significant cost differences. They analyse reasons why water privatization may prove problematic in lower-income economies but draw no conclusions - stating that the results are not significant. More poignant is the final statement, however: 'Finally it needs to be stressed that while the paper has concentrated upon a number of performance measures, a more comprehensive study would take account of possible effects beyond those discussed. For example, we have seen that privatization tends to lead to more water metering, but what is the impact of this on water consumption and health? Around major cities in developing countries lie shanty

Figure 20.8. Usual way of collecting water in Mamathe, Lesotho.

towns populated with squatters and others without legal property rights. How are their interests served by water privatization? Water privatization usually means the involvement of a handful of major international companies, but what effect does this have in terms of developing indigenous ownership of socially important assets? Also, if privatization leads to full cost recovery in water, is this outcome compatible with poverty reduction and what are the environmental implications of privatization? Clearly, water privatization raises a complex set of considerations that deserve fuller exploration than has been possible here because of data limitations' (Kirkpatrick et al., 2004, p24). It seems that some of the water war issues and huge differences opinion are exacerbated by the fact that researchers are measuring different things and, in some cases, not the ones which are relevant to issues of social justice.

One paper, which does address this directly is entitled 'Water Justice in Africa: Natural Resources Policy at the Intersection of Human Rights, Economics and Political Power' (Francis, 2005). In this paper Francis analyses water as a social justice issue in South Africa. She explores historical changes in South African water law. Francis points out that the remarkable part of South Africa is its inequality despite being post apartheid because of years of colonial and fifty years of apartheid rule. White South Africans, who comprise just 10% of the population, predominantly live in conditions similar to those in wealthy nations. However, she states that one third of the population live in conditions of less than U.S. 2$ per day with 30-60% unemployment amongst the blacks. All white suburbs have 50% of the residential water use and only 27% of black households have running water compared with 96% white. Insufficient and inadequate infrastructure for delivering water services to poor communities is compounded by scarcity in water. Much of this, Francis suggests, goes back to apartheid. Black citizens were prohibited from residing in urban areas and could only have legal residence in crowded 'homelands.' They were forced to live in areas with poor soil and limited water, to clear out unwanted shantytowns and make room for commercial forestry etc. on valuable land. First and third world economies developed independently of one another ... 'Current discrepancies in the allocation of water can be traced back to apartheid which reserved the resource for the landed white minority through convoluted ... laws linking water rights to land ownership.' Most water was used inefficiently and virtually for free by large scale commercial farmers – a 'dominant group' with 'privileged access to land, water, and economic power' (Francis, 2005, p8).

In 1990, sitting president Frederik Willem de Klerk announced the end of apartheid and released from prison Nelson Mandela, the leader of the African National Congress (ANC). In 1994, ANC came into power. Today, there is a Constitution that embraces human rights' principles and contains a Bill of Rights, which enshrines rights to many basic services including water. However, says Francis, within two years, they had transformed into neoliberal policies aimed at macroeconomic growth and restrictions on public spending – lauded by the IMF. In such a system, she explains, it is the poorest who disproportionately pay the price for economic restructuring. 'The country has not realised significant foreign investment but instead has shed hundreds of 1000s of jobs primarily in sectors like agriculture and textiles which tend to employ poor black women' (Francis, 2005, p13) 5% of the population controls 80% of the countries wealth.

The Water Services act of 1997 codified the Constitutional right to basic water and sanitation bringing piped water within 20 metres of each household. The national water act (NWA) of 1998 is widely considered to be one of the world most progressive water policies on paper, and it disconnects water rights from land ownership. The NWA was widely supported by black Africans but opposed by business interests, white farmers and outgoing National party – however, according to Francis, it creates loopholes that allow the status quo to prevail and facilitates the implementation of neoliberal financial policies.

'Insufficient access to potable water was a widespread problem before the ANC came into power, but to the extent that water was available, it was provided for free or at a highly subsidised price.' The government then introduced cost recovery and the price of water increased dramatically and the connection fees and volumetric charges proved too costly for low income households who had been getting their water from communal taps. 'As a result many individuals were forced to resort to collecting water from streams, canals or stagnant puddles' (Francis, 2005, p25).

She quotes Metolina Mthembu, a 70 year old resident in Mbabe village in KwaZulu-Natal province, 'who is glad the new government installed a tap outside her home two years ago. But now the water costs money and people here are poor. There are no jobs. We must choose between food and water, so we buy food and pray that water does not make us ill. It is a bad gamble. Many, many of us have grown sick from the water' (Francis, 2005, p25).

There is now a substantial debt among low income families due to water service charges, which according to one national survey averaged U.S. 290 by 2001. Amazingly, there seems to be a debate about why there is a non payment of water services and suggestions about a culture of non payment. Water cut offs are the most common response to inability to pay. More than 10 million have had their water disconnected. Meantime, local governments, Francis says, are trying to turn water services into profitable ventures in order to attract private investors, encouraged by the IMF and World Bank.

Challenge Box: If your city has recently experienced privatization of water, try to find out who the major shareholder is in the company that runs the organisation. Locate any news articles that you can about this company and consider the various viewpoints expressed.

20.6 PRIVATISATION

We have been discussing privatization of water and we might assume what this means. It is worth, however looking more deeply at the various definitions of this phenomenon.

Artists representation of a stunt organised by the UK National WDM (World Development Movement) and carried out by Brighton and Hove, UK World Development Movement at the 2006 Labour Party Conference. It was to highlight WDM's Dirty Aid Dirty Water campaign to protest against the UK Development organisation's use of aid money to pay private consultants to push water privatisation on developing countries rather than supporting improvement of public utility provision.

Figure 20.9. Water for sale on Brighton Beach.

'in its narrowest sense , privatization happens when the state sells its assets to a private company, along with all of the maintenance, planning and operational responsibilities that these assets entail. Over the past 30 years states have divested themselves of airlines, railroads, telephone services, health facilities and other services, thereby unlocking a new phase of capitalist expansion and innovation. Divestiture, as this form of privatization was the model of privatization adopted in the UK under Margaret Thatcher in the late 1980s with entire systems of public services delivery being sold to private firms'(McDonald and Rulters, 2006 p9)

The South Africa Workers Movement (SAMWU), says that privatization takes place when (Samson, 2003):

- The government contracts private companies to run certain parts of a service. This they call outsourcing.

- The government gets a private company to manage a government department or unit. This is usually called a management contract.

- The government gets community groups to do work that used to be done by the municipality.

- Government departments are changed into private companies which are owned by government. In Jo'burg these private companies are called utilities, agencies and corporatised entities or UACs.

- Government departments are changed into business units, which are completely separated from other departments in the municipality. These units are still owned by government but they operate like private businesses with the same kind of profit incentives (Samson, 2003, p13).

McDonald and Ruiters (2005/6) explain the different forms of public private partnership in full in Table 20.3.

Table 20.3: Different forms of public private partnership (McDonald and Ruiters, 2005/6)	
Full Divestiture	Divestiture refers to a situation where a public utility or service has been fully privatized.
Service Contract	This is the least risky of all partnership types. The public authority retains responsibility for operation and maintenance of the service - components are contracted out. 1-2 years
Management Contract	The management contractor operates and maintains services or parts of services. 2 - 20 years
Lease or Affermage	The lessor rents the facility from the public authority which transfers complete managerial responsibility for operating and maintaining the system to a private company
Concession	Investment linked contract. Concessionaire has responsibility for services and capital investments. Ownership of fixed asses is assigned to the local authority at the end of the contract. 25-30 years
BOOT	Build Own Operate and Transfer contracts - new parts of a service system. 25 years
Community/NGO Provision	Transfer of some or all of the responsibility for service provision to end user or not-for-profit intermediary. Common in low income urban settlements. Women often carry the burden of this labour.

Sometimes we confuse terms such as privatization, corporatisation and commercialisation. Macdonald helpfully teases out the differences. Commercialisation, he suggests, refers to a process by which market mechanisms and market practices are introduced into the operational decision making of a public service. A popular institutional form of commercialisation is corporatisation where services are 'ringfenced' into stand alone business units owned and operated by the state but run on market principles. The link between corporatisation and privatization is that there has been a change in the management ethos to an increasing focus on a narrow short term financial bottom line. This means that public enterprises can be even more commercial that privatized counterparts. Corporitisation also promotes outsourcing as an operating strategy and cost cutting, and it can act as a gateway for direct private sector investment.

'Tying all this together are the underlying processes of commodification. Only when public services are treated as a commodity can they be effectively commercialised and eventually privatised. It is at this politico-economic juncture that we see the full significance of such as transformation emerge e.g., water stripped of its image as an abundant nature provided good for public benefit, to water as a scarce monetised entity subject to the same laws and principles of market as shoes, lampshade or computers' (McDonald and Rulters, 2005/6, p13).

20.7   WHY IS PRIVATIZATION TAKING PLACE?

Neoliberal analysts have argued that privatization occurs because states fail: state officials are rent seeking, inefficient, unaccountable, inflexible and unimaginative. Privatisation is seen as a rational and pro poor policy choice, obvious to anyone willing to look at track records of public versus private debate. McDonald and Rulters quote Hodge (McDonald and Rulters, 2005/6, p15) who identifies five core theories: public choice theory, agency theory, translation costs analysis, new public management and property rights theory – they all use the assumption that people respond best to market incentives and that market based systems are inherently more efficient. Macdonald et al. argue by contrast 'that the privatization of public services has not happened because it has been inspired by some renewed sense of cultural enthusiasm for the market but rather that it has become a necessity imposed on the state by economic circumstances' (McDonald and Rulters, 2005/6, p16). They state that it is possible to trace the shift with water. Economies and construction of new dams shrank in 1970s and competition for opportunities intensified. The state could not support spending on new infrastructure. For the water and engineering construction industry, 'privatization was seen as a way to absorb idle productive capacity and excess commodities' (McDonald and Rulters, 2005/6, p16).

They say that nowhere in the mainstream literature is there a recognition of the argument that privatization and commercialization are a response to the pressures of an ever expanding marketisation of social relations under capitalism. Nor is there any discussion in mainstream debates of the radical thesis that capitalists must constantly seek new geographies and sectoral areas of investment as a response to a capital overaccumulation or that capitalists are constantly forced to recreate the physical means of production of built environments that facilitate market expansion. For global capital seeking new areas, they go on to say ' the public sector provides an enticing opportunity. Water is particularly attractive' (McDonald and Ruiters, 2005/6, p16).

Barlow, however, maintains that strong civil resistance is the key to the retreats that have been seen so far. 'Today a coordinated and highly effective international water justice movement is fighting both the power of the private water companies and the abandonment by their governments of the responsibility to care for their national water resources and provide clean water to their people' (Barlow, 2008, p124).

20.8   FINAL THOUGHTS

In this chapter, we have presented strong opinions for and against the issue of the privatization and commodification of water. You may already be paying for your water bills and know how expensive they can be. Hopefully, we have given you enough food for thought to allow you to consider the movements of your own government on this issue, but also to help you think through some of these perspectives if you begin to work for an engineering company who is responsible for carrying out the decisions made about this at a political level.

Challenge Box: Consider the pros and cons of privatization of public services after all the various discussions in this chapter.

C H A P T E R 21

Awakened by an Alarm Clock

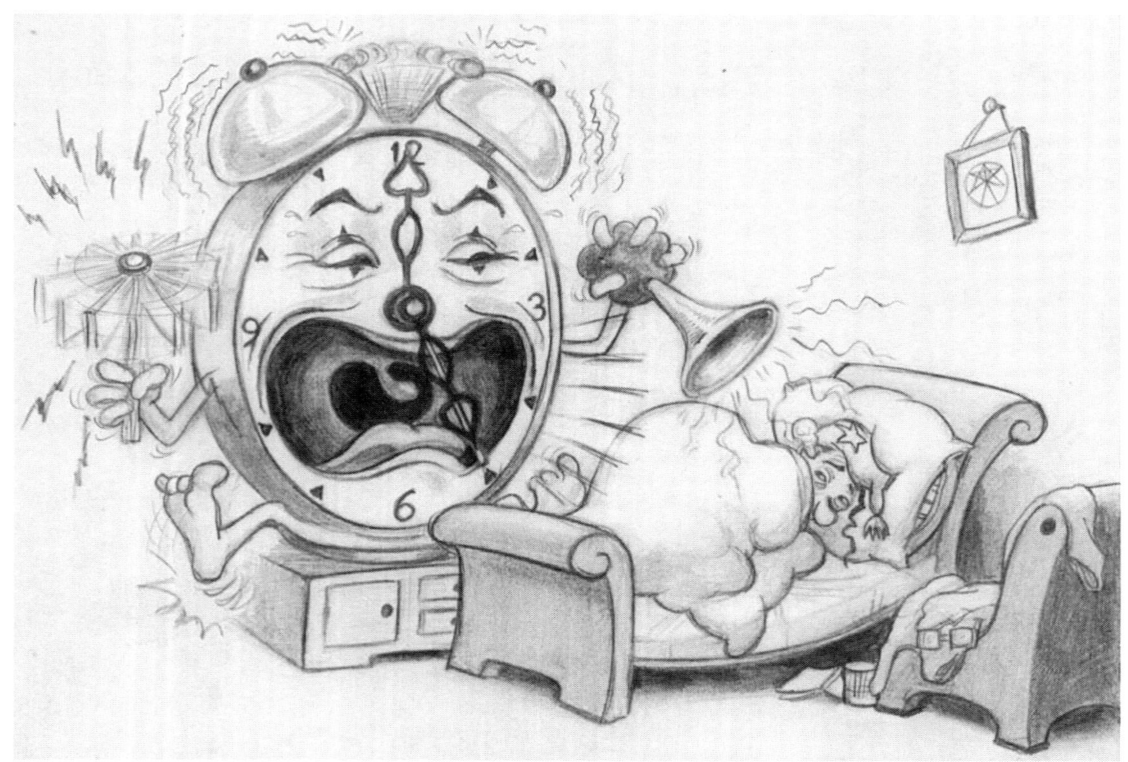

Figure 21.1. Awakened by an alarm clock.

In this chapter, we are focusing on time. As a citizen, we wait for time to pass or we want to try to beat time. As an engineer, many of our processes are controlled by time. Looking at time from the perspective of social justice causes us to ask different questions.

21.1 ENGINEERING AND OUR CONCEPT OF TIME ARE INTRICATELY RELATED

The alarm clock over there on the nightstand, nestled among the books, magazines and cough drops, that lies silently, stealthily in wait to roust you from the deepest of sleeps, is one of the most important symbols or metaphors of our modern, technologically advanced society. The clock, in many ways rules our lives, telling us when to wake up, when to get off to class, when to meet a friend, really when to do just about anything! In this next section, we will explore how modern society came to be governed by this merciless ruler and speculate on how things might be different if we adopt a new understanding of time. This will be true for all professions, but the resultant consequences for engineering may be the greatest of all.

By the end of this section it is hoped that you will be able to:

- Describe the importance that *time* plays in modern, technological society.

- Describe the links that exist between our concept of time and our most commonly held scientific paradigm, i.e., *nature as a machine.*

- Describe the relationships that exist between *nature as a machine* paradigm and other academic disciplines.

- Understand a new scientific paradigm, which has superseded our notion of nature and the universe as mechanical.

- Appreciate how the new paradigm may change our understanding of time.

- Examine the possible future implications of this new understanding for engineering.

- Elucidate possible consequences for this new understanding in other disciplines.

Time present and time past
Are both perhaps present in time future,
And time future contained in time past.
If all time is eternally present.
All time is unredeemable.
T.S. Elliot.

In *Technology and Society*, we pointed to the connections that exist between technology, engineering practice and the society in which it operates. As was discussed, engineering practice and the development of technology operate as systems of thought and practice, which provide a framework for what, we imagine and what we wish to create. Different sets of values in a society could lead to an entirely different set of technological developments.

In the present chapter, we will focus on the alarm clock, that ubiquitous and ever-present feature of modern life. More than simply a collection of gears and levers or circuit boards and LED's, the alarm clock is symbolic of many important aspects of modern life, has significance as an indicator of our societal values, and foreshadows a future for both our society and our planet. We will look back at different technological developments and societal changes that led up to the alarm clock and our conceptual model of time. We will then consider the consequences that our present model for time has in both the short term and long term. Lastly, we will imagine a very different model of time and explore the possibilities that might result.

Challenge Box: Think of five ways in which the clock governs your behavior during the day. For each way you have identified, is there some connection with an important aspect of modern society?

21.2   THE ALARM CLOCK

A **clock** (from the Latin *cloca*, "bell") is an instrument for measuring time and in its modern form dates back to approximately the 15th century. Adding the alarm to the clock changes the device from one that simply measures the quantity *time* to one that exerts control over our lives or at least attempts to do so! Today, we move from doctor's appointment to dentist's appointment, from one classroom to the next with precision that rivals that found in a highly trained military organization. When we need to travel via airplane, train or bus we consult a schedule which provides us arrival times and departure times, sometimes even accurately. We arrange our evenings and weekends around the posted times of various entertainment outlets from movies and plays at our nearby theaters to situation comedy TV shows to kickoff times of Super Bowls and World Cups. On some occasions, you may even wish to consult with that favorite professor during her/his announced office hours with a reliance again on the universality of *time*. Everything, it seems is governed by what we have come to regard as the phenomenon known as *time*.

As engineers, the present day understanding of time is integral to the practice of our profession. Engineers model a wide range of phenomena, postulating equations, which predict behavior. We describe the resulting solutions as either steady state or unsteady or transient, depending upon whether those solutions are independent of time, change with time without ever reaching a steady state value, or change with time until eventually a steady state value is reached. In fact, engineers today think of *time* as a fourth independent variable, comparable to the three independent variables we typically associate with three-dimensional space. As engineers, we are often tasked with solving problems that have never been solved before using our design methodologies, integrating analytical and creative thinking skills. In design, we develop milestone schedules to layout and track the various tasks that will be required to reach the desired end goal. Whatever design strategy we chose, it is time that serves as the underlying concern. Our final design and product must be delivered at a certain specified and agreed upon time.

It is the clock that tells us the appropriate time. Yet, the clock, particularly the mechanical clock with or without the bells and whistles necessary to sound the alarm, has only been in widespread use for the last several centuries. We shall first consider the importance of time in present day society. Then in this next section, we will look back in time to consider one important technological advancement, the mechanical clock, that leads up to our conception of time today. We shall then imagine the role of time in the future and consider the implications of our present understanding for that possible future.

21.3   MODERN SOCIETY AND WELCOME TO THE MACHINE

> Welcome my son, welcome to the machine.
> Where have you been? It's alright we know where you've been.
> You've been in the pipeline, filling in time, provided with toys and Scouting for Boys.
> You bought a guitar to punish your ma,
> And you didn't like school, and you know you're nobody's fool,
> So welcome to the machine. (Pink Floyd, *Welcome to the Machine.*)

Modern engineering has been intricately linked to the design, development and manufacture of machines. As a society, we expect our machines to run as "clocks," that is, we wish our designed products to operate dependably, with repeatability and preciseness. The clock is often used as a metaphor for modern industrialized society, and until very recently, we were often times described as living in the "machine age." Our modern, highly mechanized civilization is the outgrowth of the design and development of countless engineering devices as well as a combination of numerous habits, ideas customs and cultural norms. Even the universe itself is considered by many in technological fields to be the ultimate machine whose behavior is understandable and thus predictable *when*, not *if*, science finally discovers the proper set of equations or as physicists refer to it, the *theory of everything* (Hawking, 1998).

As you walk about the common grounds of your college campuses, you are likely to hear bell chimes emanating from the campus campanile or clock tower. Those tones tell you the time of day and whether or not to walk leisurely or sprint at top speed to your next class. The universally recognizable pattern of tones breaks up the day into twenty-four hours and typically each hour into fourths or 15-minute intervals. The bell towers date back to the 13th and 14th centuries in Western Europe, the most famous being the clock built by Heinrich von Wyck in Paris. With the advent of the clock and its ever imposing and obvious presence, time soon took on a very different character then it had earlier. According to Mumford, "Time took on the character of an enclosed space: it could be divided, it could be filled up, it could even be expanded by the invention of labor-saving instruments" (Mumford, 1963). Engineering soon became tasked with designing, developing and delivering instruments or devices, which would enable the user to save time and thereby do more.

Engineers also became concerned with designing and implementing process that would save time as well leading to the field of time and motion studies or time-motion study. A time and motion study is used to develop a new process or series of sequences/steps to reduce the number of motions in performing a task in and increase productivity. The best known experiment involved bricklaying. Through carefully scrutinizing a bricklayer's job, Gilberth in the 19th century reduced the number of motions in laying a brick from 18 to about 5. The focus was on laying more bricks in the same amount of time and thereby expanding the time, bringing mechanical efficiency through coordination and careful design.

In many cultures, particularly in the West, efficiency was soon seen as an important and highly desirable trait. This attribute, the ability and the desire to do more and more in the same period of time, is profoundly evident today. While only a few years ago, it was sufficient to drive a car from one place to the next, now we find it necessary to talk on cell phones, text messages, use instant messenger, and countless other tasks all from behind the steering wheel. Restaurants are no longer settings for meals and casual or intimate conversations but rather a place to eat quickly, and catch all that is happening in the world while glancing at monitors tuned to stations focused on entertainment, the weather, news and sports. Our homes, dormitories, offices are all filled with all manner of gadgets and devices (all designed, developed and delivered by engineers) that are sold to us as items that will allow us pack more and more into the same 24 hour period. Perhaps, the motto that best describes the modern world is this: Just give me more time! I need more time! (Tolle, 2004).

21.4   THE ASSEMBLY LINE

An examination of modern *assembly lines* provides insight into how important the concept of time has become. An *assembly line* is a manufacturing process in which interchangeable parts are added to a product in a sequential manner to create an end product. Until the 19th century, a single craftsman or team of craftsmen would create each part of a product individually, and assemble them together into a single item, making changes in the parts so that they would fit together. This linear assembly process, or assembly line, allowed relatively unskilled laborers to add simple parts to a product. While originally not of the quality found in hand-made units, designs using an assembly line process required much less training of the assemblers, and, therefore, could be created for a lower cost (Mumford, 1971). Modern assembly lines often have much more complicated interdependencies with not only workers and but also robots and other devices linked in the desire to produce more in less time at a lower cost. Henry Ford was one of the first to apply assembly line manufacturing to the mass production of affordableautomobiles. This achievement not only revolutionized industrial production in the United States and the rest of the world, but also had such tremendous influence over modern culture that many social theorists identify this phase of economic and social history as "Fordism" (Watts, 2005).

Figure 21.2. Assembly line in Stolen water company.

An assembly line is a manufacturing process in which interchangeable parts are added to a product in a sequencial manner to create a finished product. Hounshell (1984)

While some scholars credit Henry Ford with the creation of the middle class in the United States, soon other consequences of the assembly line became apparent and still exist today. Many workers soon became unhappy with the assembly line. They felt alienated from the products of their work as they often never had the satisfaction of seeing the finished product. Workers in assembly lines can easily be thought of as simply another cog in the production machine and easily replaced. Because workers had to stand in the same place for hours and repeat the same motion hundreds of times per day, they often suffered from what are now called repetitive stress injuries. In addition, workers were often times exploited and forced to work in dehumanizing conditions.

> **Challenge Box:** Think of any assembly lines you encountered today. What was your reaction? Imagine you were working in an assembly line of some kind. How do you think you would react?

21.5   NATURE AS MACHINE

Not only has the production of goods and the role of workers in that production been seen to be as a *machine* but so too has our conception of the natural environment. Our environment today is seen by many to be a collection of *natural resources* which we can use to satisfy our needs and desires. We also speak of maintaining the balance of nature or returning nature to its balance. We imagine we can do this by adjusting a whole range of input parameters and boundary conditions much the same way we design assembly lines to maximize product output. Nature is seen as governed by the laws of *causal determinism* (Bunge, 1963) in this view.

> Causal determinism is based upon the belief that every effect has a cause and thus science, if clever enough and developed enough, can be used to explain all natural phenomena.
> The stanford Dictionary of Philosophy, 2009

> Even if there is only one possible unified theory, it is just a set of rules and equations. What is it that breathes fire into the equations and makes a universe for them to describe? The usual approach of science of constructing a mathematical model cannot answer the question of why there should be a universe for the model to describe. Why does the universe go to all the bother of existing?

In causal determinism, the only thing that can be said to exist is matter and all things are composed of matter, and all phenomena are the result of matters' interactions. Our role as engineers then is to use this information discovered by scientists to help us to design devices and/or processes that meet our clients' needs. Ultimately, we envision nature or the universe to be a machine and we are the master mechanics whose goal is to first conquer and then modify the environment to suit our interests. Greene (1999).

Challenge Box: Do you share the view that the universe is material or matter only? What makes you feel confident in your view? How did you arrive at your view?

The consequences of the *Nature as machine* model (Botkin, 1993) have become more readily apparent than ever before. Plant and animal species are becoming extinct at a rate unsurpassed at any other time except for the mass distinction associated with the dinosaur disappearance. The oceans of the world are succumbing to years upon years of over fishing and pollution of various guises (Mastny, 2005). The Earth's climate is changing with the average land and ocean surface temperatures rising faster than ever before in recorded history. This phenomenon, referred to as *Global Warming*, confronts us with serious questions about the sustainability of our life styles and through extension ultimately our personal value system (World Watch Institute, 2006).

> "An increasing body of observations gives a collective picture of a warming world and other changes in the climate system."
> Intergovernmental Panel on Climate Change - IPCC (2001).

There are many other consequences that arise from the adoption of nature as a machine model. Some of these consequences are more explicit than others but all are equally directly resulting from the mechanistic conceptualization of nature. The suggestion that nature is a machine also suggests two additional points. First, some of us, specifically engineers, design machines to do our bidding. If nature is a machine then we can certainly manipulate (or at least attempt to do so!) the forces of nature to meet our needs and desires. We often speak of conquering and/or taming nature. While no mechanist would suggest that at the present moment in history, we can conquer/tame hurricanes, we do feel dangerously confident about having our wishes. We need look no farther than the devastation wrought by Hurricane Katrina, which struck the Gulf Coast of the U.S. in 2005, or the tsunami (Unesco, 2005) that devastated nations from Southeast Asia to the eastern coast of Africa with thousands of lives lost.

> **Challenge Box:** What is your reaction to the conceptualization of nature and the universe as a machine?

21.6 ECONOMICS AND THE MACHINE

Once the "common sense" view that nature and the universe are deterministic and governed by a set of unchangeable laws, it was not long before other academic disciplines outside of the pure sciences were modeled in similar ways. This was particularly true in the field of economics and, not surprisingly, the *mechanical* view of economics still dominates today. The economist equivalent to the natural laws of physics put forward by Newton and his contemporaries is the *Law of Supply and Demand* (Smith, 2003). The law, originally developed by Marshall (1997), attempts to describe, explain and predict changes in the price and quantity of goods bought and sold in competitive markets. The law remains as a basic building block for a wide range of modern, more detailed economic theories and is an explanation of the mechanism by which many resource allocations are made. Note the use of the word *mechanism*!!

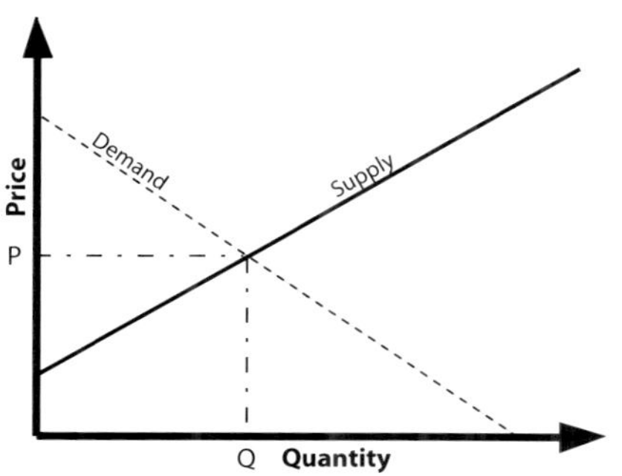

If the natural laws of physics gave us the mechanical universe, what then did the laws of economics give us? Most likely, the answer would be capitalism. Commonly, the concept of capitalism refers to an economic system in which the means of production are primarily privately owned and operated for profit, with the investment of money or capital also determined privately, and decisions regarding the production, distribution and prices of goods, services and labor influenced the laws such as the law of supply and demand. As mechanists refer to the behavior of the natural world in harmony [i.e., reminiscent of *the music of the spheres* in physics (James, 1993)] with immutable natural laws, capitalists point to "the natural harmony of the rational self-interests of all men under capitalism – of businessmen and wage earners, of consumers and producers, of men of all races and of the competition of all levels."

21.7 LOOKING FORWARD IN TIME: A NEW PARADIGM FOR SCIENCE AND TECHNOLOGY

Science no longer conceptualizes nature and the universe as a machine with predictable patterns of behavior in the same manner as a mechanical clock. Today, we speak of self-organizing principles, and emergent properties (Bak, 1996). Self-organizing principles govern a process in which the internal organization of a system, normally an open system, improves automatically without being guided or managed by an outside source. Self-organizing systems typically display emergent properties. Emergent properties are properties that are dynamic in nature, that is, they change in time, and grow in complexity. Ultimately, the resultant structures dissipate or dissolve in time and the process begins anew. An emergent property can appear when a number of simple entities or agents operate in an environment, forming more complex behaviors as a collective. Examples of such properties include the construction of an ant hill or a bee hive wherein ants or bees come together and build complex structures that we as engineers often seek to emulate! Nature is thus not a collection of pieces and parts that can be replaced at our discretion. Once nature starts a process or we impact the starting of that process, we have no control over its final disposition. We no longer can speak of taming or conquering the natural world. That simply isn't in the cards!

Implicit in the new paradigm is a greater reliance upon community. It is not the individual ant or bee that matters but the entire population of ants and bees. A second implication of this new paradigm is that very simple actions ultimately may lead to very complicated and unforeseen results. Perhaps nature plays its own version of the *Rube Goldberg* game (Wolfe, 2000).

Figure 21.3. Rube Goldberg machine.

Rube Goldberg gets his think-tank working and evolves the simplified pencil-sharpener. Open window **(A)** and fly kite **(B)**. String **(C)** lifts small door **(D)** allowing moths **(E)** to escape and eat red flannel shirt **(F)**. As weight of shirt becomes less, shoe **(G)** steps on switch **(H)** which heats electric iron **(I)** and burns hole in pants **(J)**. Smoke **(K)** enters hole in tree **(L)**, smoking out opossum **(M)** which jumps into basket **(N)**, pulling rope **(O)** and lifting cage **(P)**, allowing woodpecker **(Q)** to chew wood from pencil **(R)**, exposing lead. Emergency knife **(S)** is always handy in case opossum or the woodpecker gets sick and can't work.

21.8 SELF GOVERNING PRINCIPLES, ECONOMICS, AND COMMUNITY

Self-organization is also relevant in chemistry, and is central to the description of biological systems, from the sub-cellular to the ecosystem level. Self-organizing behavior found in the literature of many other disciplines, both in the natural sciences and the socila sciences such as economics or anthropology.

Figure 21.4. Tennyson (2009)

In economics, according to some economists, the theoretical free market is self-organizing. Other economists have offered a completely different view. Bogdanov, scientist, philosopher, economist, physician, novelist, poet, and Marxist revolutionary, suggested that all human, biological and physical activities can be unified, by considering them as systems of relationships, and by seeking the organizational principles that underlie all kind of systems (Biggart et al., 1998). According to Bogdanov, it is fundamentally wrong to separate out human activity from biological evolution and physical phenomena, and any such effort would result in significantly harsh consequences for our planet.

A second consequence of the self-organizing results from the importance of community whether it is the community of ants, bees or in a professional setting, a community of workers. The whole is not simply the sum of all the individual pieces and parts but something quite different.

This is the duty of our generation as we enter the 21st century – solidarity with the weak, the persecuted, the lonely, the sick, and those in despair. It is expressed by the desire to give a noble and humanizing meaning to a community in which all members will define themselves not by their own identity but by that of others.

Challenge Box: How do you react to the model of the universe as a community of interests rather than a collection of pieces and parts? How does that effect your understanding of the responsabilities of engineering?

21.9   TIME IN A MACHINE; TIME IN A COMMUNITY

Mechanical devices, assembly lines, natural laws of science and economics—they all point to the notion of the linear progression of time. Things proceed in an orderly, quantifiable way. We can identify initial conditions and boundary conditions and watch with delight as the process goes forward. In addition, we have a sense of being able to control the progression. We can twist knobs; fiddle with initial and boundary conditions to arrive at the optimum design. Take the mechanical clock for example. If we wish to slow it down, we can extend the length of the swinging pendulum or change the mass of the bob. On an assembly line, we can change the order of the parts being assembled or the workers who are employed. With the proper modifications, we can achieve whatever result we have decided to pursue. If we wish to build a resort city in the desert, say for example, Las Vegas, we can do just that. If we wish to prevent a river, for example the mighty Mississippi, from changing its path to the sea, we can do that as well for as long as we wish. With each technological advance, there is a sense that progress continues and with enough time we will be able to solve any and all problems that presently confront us. Such a view of the world seems very comforting, yet it is the product of a science which has long since been replaced.

Suppose an Objective Observer were to measure the success of Progress - that is to say, the capital-P myth that ever since the Enlightenment has nurtured and guided and presided over the happy marriage of science and capitalism that has produced modern industrial civilization.

Has it been, on the whole, better or worse for the human species? Other species? Has it brought humans more happiness than there was before? More justice? More equality? More efficiency? And if its ends have proven to be more benign than not, what of its means? At what price have its benefits been won? And are they sustainable?

With a shifting scientific paradigm, substituting the science of community for the science of mechanics, the physical property, time, remains the same but its significance for us in trying to understand natural processes and, in fact, our place in the universe is very different. Rather than speaking of a universal time, we would speak of an entire spectrum of time scales from incredibly small to the magnitudes associated with the size of the universe. In addition, rather than projecting time forward as a straight and true arrow, we would reflect upon the cyclical behavior of many phenomena and use the geometric shape, a circle, as a metaphor for that behavior.

Time depicted as a circle rather than an arrow decouples the passage of time from the myth of progress. With each new advance, we do not know where we are going ultimately and hence we cannot argue that the present will be necessarily better than the past or the present. It may or may not be depending in part on how we go about designing and implementing our proposed solutions.

21.10 CONCLUSIONS

Buzz! The alarm clock just went off so we need to move on to the next thing scheduled and then the next and the next and so on *ad infinitum*! We explored the historical origin of our present conceptualization of time and discovered it is linked to a scientific paradigm that describes the universe as a mechanical device or machine. We also have discovered that science has changed and, of course, will continue to change. Nature now is most often conceived as a self-organizing system or a community of interests.

You, as engineers and technologists of the future, will be charged with developing solutions to new and ever more difficult problems. As future engineers, you can choose to model your world as a mechanical clock, treating all of nature including the animal and plant worlds as well as fellow human beings as collections of pieces and parts. More is always better, and the future can be controlled to meet our needs and satisfy our desires. Alternatively, you can choose to model your world as a community of interests, each with their own needs, shortcomings, hope and fears. The former world-view has led to inestimable wealth for a few and many unanswered questions concerning its sustainability in the long run. The latter approach represents a significant change in the ways we go about our professions. It suggests that more may be a recipe for disaster and we will never have certainty over the course of events, the impact of our engineering decisions. Above all else, it cautions for deliberateness and against arrogance.

Challenge Box: How does engineering change if you consider the importance of community in you work? Who is in your community? How far do the wall of your community extend?

C H A P T E R   22

Driving the SUV

Figure 22.1. A green car?

"Human activities are increasingly altering the Earth's climate.... It is virtually certain that increasing atmospheric concentrations of carbon dioxide and other greenhouse gases will cause global surface climate to be warmer. The unprecedented increases in greenhouse gas concentrations, together with other human influences on climate over the past century and those anticipated for the future constitute a real basis for concern."
American Geophysical Union, 2003.

"Even though cars get worse gas mileage than two decades ago, they actually have become much more efficient. The problem is that the efficiency went into more power and larger, heavier vehicles, not into fuel economy. If all of the technological efficiency improvements had gone into efficiency, miles per gallon would be significantly higher than they are today."
Neal Elliott, American Council for an Energy Efficient Economy.

22.1   SO WHAT EXACTLY IS AN SUV?

SUV is the widely used and known acronym for 'sport utility vehicle.' At first conception, an SUV was a vehicle that combined the towing capacity of a full-size truck with the passenger and storage capacity of a minivan. However, as consumer demands have changed, so has the SUV. Many manufacturers now focus on fuel-efficiency and driving and riding comfort, rather than towing capacity.

Figure 22.2. An Example of an SUV (2008).

Typical features of an SUV include seating for five to seven, high seating and road positioning, roomy interior, non-dedicated trunk space, high engine capacity and 4-wheel drive capability. Though the SUV was originally designed to be an off-road vehicle for sporting purposes, their popularity has spawned several different breeds, including the luxury SUV.

As is the case for passenger cars, the SUV has different classes and sizes.

Though many modern day motorists value the SUV for its size and roominess, many others criticize their lack of fuel efficiency and their contribution to air pollution. Consumers who value the SUV do so not only for its size, but the perceived safety of driving such a bulky vehicle. Though crash test safety ratings vary with makes and models, some SUVs are known to pose the risk of rollover.

For those who enjoy the flexibility of combining weekday comfort with weekend fun, the SUV has proven to be a leading choice in vehicles. Similarly, having 4-wheel drive capability and towing capacity without the passenger restrictions of a truck appeals to many car buyers and keeps the SUV a leading seller.

As a concern for the environment and the demand for fuel-efficient vehicles increases, manufacturers worldwide are continuing to explore ways to make improvements to the SUV family to keep them a leading selling vehicle.

22.2   THE SUV AND THE ENVIRONMENT

SUVs represent a paradox to consumers - television advertisements present them as a way to return to nature, yet they actually accelerate existing environmental problems. Commercials often depict happy families driving on mountain roads, avoiding falling rocks and enjoying the flowered wilderness in leather-seated comfort. The sad truth is that these vehicles are contributing to the destruction of our natural resources. In reality, only 5 percent of SUVs are ever taken off-road, and the vast majority of these vehicles are used for everyday driving. In 1985, SUVs accounted for only 2 percent of new vehicle sales. SUVs now account for one in four new vehicles sold, and sales continued to climb until 2008.

Driving an SUV has a much greater impact on the environment than driving other passenger cars due to vastly different standards set by law and government regulations. For example, current federal regulations allow SUVs to have far worse fuel economy than other vehicles. The federal corporate average fuel economy (CAFE) standards set the fuel economy goals for new passenger cars at 27.5 miles per gallon (mpg). But under the law, SUVs are not considered cars - they are characterized as light trucks. Light trucks only have to achieve 20.7 mpg. It should be noted that this is an average for all light trucks, which is why it is possible to have SUVs on the road that only achieve 12 mpg. When CAFE was instituted in the 1970s, there were few SUVs and light trucks on the road, and they were primarily used for farm and commercial work. Today, however, the demographics of an SUV buyer are quite different. The amount of gasoline burned by a vehicle is important for several reasons. The most crucial is the threat of global warming.

22.3   THE THREAT FROM GLOBAL WARMING

Global warming has been extremely well studied. In 2001, the Intergovernmental Panel on Climate Change (IPCC) issued a report on global warming with many dire predictions. The World Meteorological Organization and the United Nations Environment Programme created the IPCC in 1988 to study the risks associated with global climate change. The IPCC found that about three quarters of the anthropogenic (caused by humans) emissions of carbon dioxide to the atmosphere during the past 20 years is due to fossil fuel burning. The IPCC anticipates higher temperatures and heat waves over the next century, as well as more intense and dangerous storms.

According to the EPA (Environmental Protection Agency), "increasing concentrations of greenhouse gases are likely to accelerate the rate of climate change. Scientists expect that the average global surface temperature may rise 1-4.5°F (0.6-2.5°C) in the next fifty years, and 2.2-10°F (1.4-

Figure 22.3. Contributing to global warming.

5.8°C) in the next century, with significant regional variation. Evaporation will increase as the climate warms, which will increase average global precipitation. Soil moisture is likely to decline in many regions, and intense rainstorms are likely to become more frequent. Sea level is likely to rise two feet along most of the U.S. coast."

The first reliable global measurements of temperature from NASA, published by Hansen and his colleagues in 1981, showed a modest warming from 1880 to 1980, with only a slight dip in temperatures from 1940 to 1970 [Graph adapted from Hansen et al. (1981)].

In 1981, NASA scientists predicted the impact of carbon dioxide emissions on global temperatures between 1950 and 2100 based on different scenarios for energy growth rates and energy source. If energy use stayed constant at 1980 levels (scenario 3, bottom lines), temperatures were predicted to rise just over 1°C. If energy use grew moderately (scenario 2, middle lines), warming would be 1–2.5 °C. Fast growth (scenario 1, top lines) would cause 3–4°C of warming. In each scenario, the warming was predicted to be less if some of the energy was supplied by non-fossil (re-

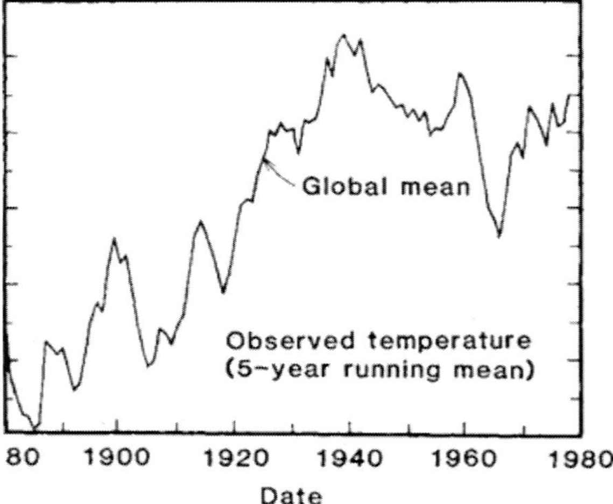

Figure 22.4. Historical record of global mean temperature.

newable) fuels instead of coal-based, synthetic fuels (synfuels) [Graph adapted from Hansen et al. (1981)].

According to the EPA, one of the most important things you can do to reduce global warming pollution is to buy a vehicle with higher fuel economy. Every gallon of gasoline your vehicle burns puts 20 pounds of carbon dioxide (CO_2) into the atmosphere. Scientific evidence strongly suggests that the rapid buildup of CO_2 and other greenhouse gases in the atmosphere is raising the earth's temperature and changing the earth's climate with potentially serious consequences. Choosing a vehicle that gets 25 rather than 20 miles per gallon will prevent 10 tons of CO_2 from being released over the lifetime of your vehicle. Passenger cars and trucks account for about 20 percent of all U.S. CO_2 emissions.

Note that a vehicle which gets 10 mpg emits approximately 120 tons of CO_2 while one that gets 30 mpg emits less than half that amount. The National Academy of Sciences estimates that if fuel economy had not been improved in the late 1970s, U.S. fuel consumption would be about 2.8 million barrels of oil per day higher than it is. Unfortunately fleet-wide improvements in vehicle fuel economy occurred from the middle 1970s through the late 1980s. Since then fuel economy has been consistently falling. In fact, average new vehicle fuel economy fell in 2000 to 24 mpg, its lowest level 20 years primarily due to the increase in numbers of SUVs.

For more information, the EPA has a website devoted to global warming, which answers many of the common questions about the problem.
(http://www.epa.gov/globalwarming/).

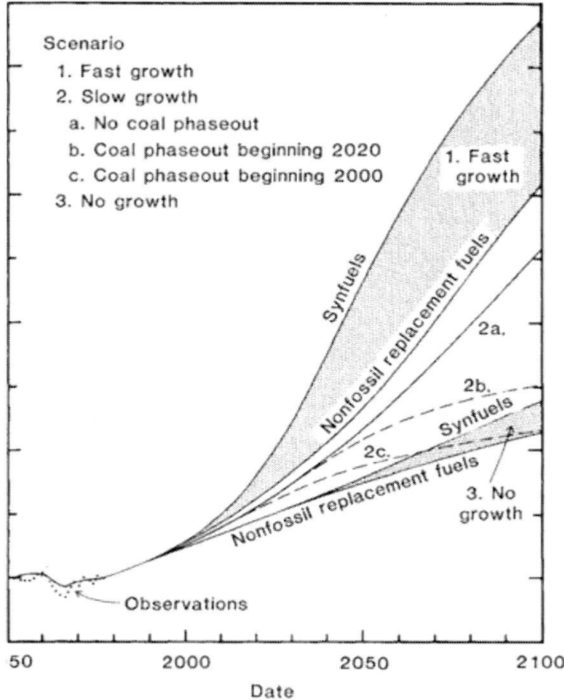

Figure 22.5. Growth of CO_2 for various scenarios.

22.4 SUV'S SMOG FORMING EMISSIONS

SUV's have a significant environmental impact even beyond the problem of global warming. Federal law gives heavy sport utility vehicles permission to emit higher levels of toxic and noxious pollution - carbon monoxide, hydrocarbons, and nitrogen oxides. Sport utility vehicles emit 30 percent more carbon monoxide and hydrocarbons and 75 percent more nitrogen oxides than passenger cars. These combustion pollutants contribute to eye and throat irritation, coughing, nausea, dizziness, fatigue, confusion and headaches. Hydrocarbons and nitrogen oxides are precursors to ground level ozone, which causes asthma and lung damage. Unfortunately, increasing numbers of Americans are living in areas with poor air quality from ozone pollution, according to the American Lung Association (ALA). The ALA found that 141 million Americans lived in areas with poor air quality during 1997-1999, an increase of nine million since 1997-1999.

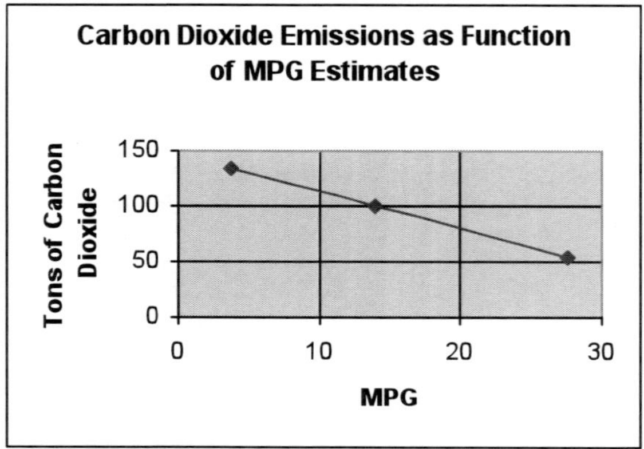

Figure 22.6. Tons of carbon dioxide emitted over vehicle lifetime.

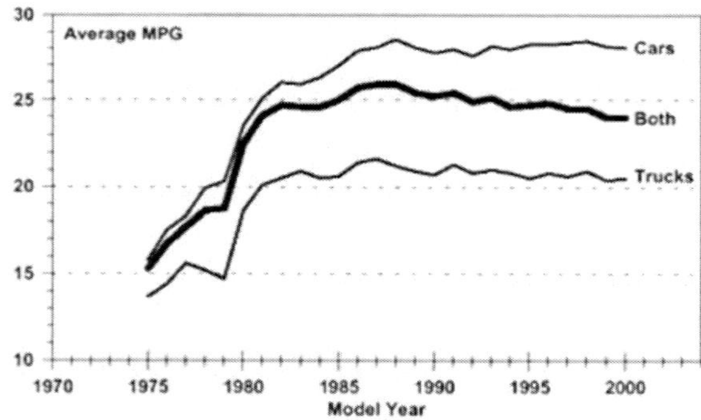

Figure 22.7. Fleet-wide Fuel Economy from 1975 to 2000.

22.5   U.S. DEPENDENCY ON OIL

Finally, it is important to note that SUVs are contributing to our dependence on imported oil. The more gasoline we use, the more oil we have to import from other countries. Currently, more than half of the oil we use is imported.

Figure 22.8 depicts the increase in the importation of foreign oil over the course of the last half century. Note that in 1993, the U.S. imported more oil than it produced and that trend has grown continuously.

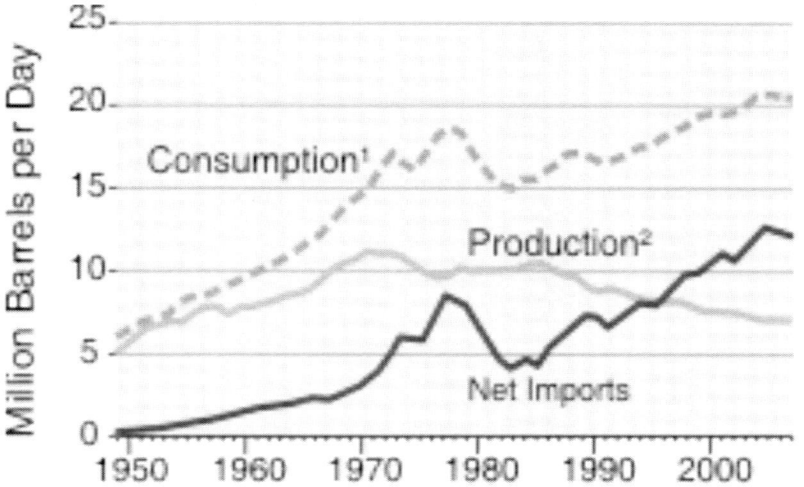

Figure 22.8. Production, Consumption and Importation of Oil in the U.S.

22.6   WHAT CAN BE DONE?

In July 2001, the National Academy of Sciences (NAS) released a study on fuel economy standards. The NAS found that light trucks, SUVs, minivans, and pickup trucks could reach 28-30 mpg for an additional cost of $1,200-$1,300. The Union of Concerned Scientists (UCS) has also looked at this issue, and found similar results. The UCS report concluded similar fuel economy levels were achievable at nearly identical consumer cost - all using existing technologies that automakers could implement quickly.

Buying a sport utility vehicle can be tempting but it's important to understand that it may mean extra expenses. Automakers have been able to charge high prices for sport utility vehicles (SUVs) making them very profitable. At the beginning of 2008, the light trucks - including SUVs - now make about half of vehicle sales in the U.S. But light trucks provide about two-thirds of profits for the Big Three automakers. While U.S. automakers make little or no profit on other passenger cars, a large SUV can bring $12,000 to $20,000 in SUVs often have heavier maintenance costs, greater gasoline bills, and higher insurance rates than other cars. In addition, recent research shows that even minor accidents can result in tremendous repair costs. The reason that the SUVs cost so much to repair has to do with federal regulations. Passenger car bumpers have to meet federal standards in low-speed crashes, and most of the bumpers on passenger cars include a reinforced bumper bar and foam to absorb crash energy. But SUVs are not subject to any kind of bumper requirements, so they are allowed to crumble in low-speed accidents.

More generally, taxpayers are also bearing an increased burden from the growth in SUV sales. This is due to a loophole in the federal tax code that exempts SUVs from a tax imposed on other cars. Known as the "gas guzzler tax" - it was instituted in 1978 to encourage automakers to manufacture fuel-efficient vehicles. More efficient vehicles pay little or no tax. Less efficient vehicles pay more tax. Except for SUVs which pay no tax. Several environmental groups have estimated that this loophole costs approximately $1.1 billion in revenues that must be made up by the typical taxpayer. As a result, we all pay more taxes while automakers produce more polluting vehicles.

22.7   CASE STUDY

Scenario: Market analysis has shown that there is a rapidly increasing demand for SUVs in various third-world countries. Several large multi-national corporations have decided to first import and then locate manufacturing and assembly plants for the SUVs in the various countries. The SUV is seen as an important element in the economic development of the countries by the ruling governments. You are a design engineer for the largest corporation and are charged with developing an 'appropriate' design and prototype of the proposed vehicle.

Approach: Using the various decision making models described in Book 2, let us explore the kinds of questions that such models challenge us to include in our deliberations and ultimately our design.

- Engineering and Freedom

Consider first an engineering decision model based upon freedom. We considered three kinds of freedom: existential freedom, substantial freedom and the freedom of choices. According to Sartre, that is, freedom as constructed from an existentialist perspective, must take on the responsibility of choosing for all of humankind, desire and work for the freedom of all humanity, and create ourselves within the context of the relationships and obligations we have to others. In describing substantial freedom Sen states that "the perspective of freedom" is concerned with "enhancing the lives we lead and the freedoms we enjoy." The ethic of freedom calls for, "expanding the freedoms we have reason to value," so that our lives will be "richer and more unfettered" and we will be able to become "fuller social persons, exercising our own volitions and interacting with–and influencing–the world in which we live." Nussbaum describes, "a twofold intuition about human beings: namely, that all, just by being human, are of equal dignity and worth, no matter where they are situated in society, and that the primary source of this worth is a power of moral choice within them, a power that consists in the ability to plan a life in accordance with one's own evaluation of ends."

So what kinds of questions should we consider in our SUV design deliberations? Sartre might suggest that we consider if we would want more SUVs here on our roads. Can we accept more congestion, more pollution, more noise and more dependence on fossil fuels? Sen might challenge us to consider whether or not more SUVs in the various countries expand the freedoms of the native populations at all? Are the citizens more able to live their lives or are their lives more restrained or constrained by the presence of the new vehicles? Both might ask us to fully explore the notion of economic development. What does that term actually mean and who benefits from such development? Nussbaum might caution us to consider whether or not all members of society are considered to have "equal moral worth" and ask whether or not the SUV promotes or at a minimum does not prevent citizens from living a life consistent with their own evaluations.

- Engineering and Chaos

Our understanding of chaos led us to the following prescriptive principle for making engineering decisions:

A thing is right when it tends to allow the natural world and all the entities thereof, to thrive in richness and diversity, and to experience change. It is wrong when it tends otherwise.

Engineering decisions using a chaos perspective caution us to focus on the health of the various ecosystems and the natural world in general. What are the impacts of the SUVs on the ecosystems were they are produced? What about where they are used? What are the short term impacts? What are the longer term impacts? What effects will the increase in SUVs have on the various cultures–the culture in which the SUV is produced and that culture in which more and more are put on the road?

- Engineering and a Morally Deep World

A decision model based on a morally deep world suggests a self-organizing system (our planet) is characterized by synthesis rather than analysis and suggests a new code of responsibility based upon community rather than individuality. For a morally deep world, the first fundamental canon and rule of practice is specified as:

Engineers, in the fulfillment of their professional duties, shall hold paramount the safety, health and welfare of the identified integral community.

The fundamental difference between an ethical code based on a morally deep world versus our present sense of responsibility is the replacement of the "public" by the "identified integral community." Who then is included in the integral community?

- Multinational corporations? Their shareholders? Their employees?
- People who will drive the SUVs? Those that will service them?
- The ecosystems that surround the various production facilities? The ecosystems that are affected by the shipping of the SUVs?
- The indigenous communities in which they will be sold? Their customs and ways of life?
- Anyone else?

- Engineering and Globalism

As we discussed earlier, globalization in its literal sense is the process of making, transformation of some things or phenomena into global ones. It can be described as a process by which the people of the world are unified into a single society and function together. This process is a combination of economic, technological, socio-cultural and political forces. The question that we are asked to consider is the following: Is the production and use of more SUVs consistent or in opposition to the *Global Ethic*? By *Global Ethic* is meant the necessary minimum of common values, standards and basic attitudes or alternatively a minimal basic consensus relating to binding values, irrevocable standards and moral attitudes.

- Engineering and Love

In order to understand the nature of an engineering based on love, we sought wisdom from Thomas Berry. Berry's most famous quotation is:

The Universe and thus the Earth is a communion of subjects, not a collection of objects.

By communion, Berry was referring to intimacy or a feeling of emotional closeness, a connection, especially one in which something is communicated or shared. By subject, the reference is to the essential nature or substance of something as distinguished from its attributes. According to Berry, our new community is a very special one; that is, it is one in which the various elements are bound together as subjects having interests rather than one in which some have interests while others are simply resources to be utilized. What are the implications then for our SUV project? Have we considered those who build, those who ship, those who buy, and those who will use our SUV as objects or as subjects? Are they simply cogs in our wheels of career advancement? Economic development? And what of the various ecosytems? How will they be impacted?

Travelling to Waikiki Beach

Figure 23.1. Going to the beach with the kitchen sink.

The native language is soft and liquid and flexible and in every way efficient and satisfactory—till you get mad; then there you are; there isn't anything in it to swear with. Good judges all say it is the best Sunday language there is. But then all the other six days in the week it just hangs idle on your hands; it isn't any good for business and you can't work a telephone with it. Many a time the attention of the missionaries has been called to this defect, and they are always promising they are going to fix it; but no, they go fooling along and fooling along and nothing is done.

Mark Twain's Speeches, 1923 ed. "Welcome Home."

23.1 INTRODUCTION

Mention Waikiki Beach in Honolulu, Hawaii, and we immediately think of beautiful white beaches, breathtaking panoramas, waves that meet the needs of even the most adventuresome surfer, and throngs of tourists parading along the sea-side promenade, thousands more busily going from exclusive boutique to novelty store, arms loaded down with countless boxes and shopping bags. Gaze out to sea and there is the magnificent Diamond Head peak looming ominously above the morning mists. Everywhere one looks there are the beautiful people of wealth and high fashion. Certainly, Waikiki Beach is nothing less than paradise here on Earth or is it? What about the people who inhabited the island long before Captain Cook landed?

Figure 23.2. Diamond Head as seen from Waikiki Beach.

Technological advancements have played an important role in rapid economic development and huge growth of tourism on the islands we refer to as the State of Hawaii. We shall examine both the environmental as well as societal impacts. Rapid advances in technology have had profound impacts on the Hawaiian people. We shall attempt to examine those impacts in light of many of the ideas we discussed for making engineering decisions in the 21st century (Book 2).

23.2 HISTORY

Polynesians, ancestors of the Hawaiian people, undertook the long ocean voyage from the Marquesas Islands to Hawaii at least 1,700 years ago. At European contact in 1778, an estimated 400,000 to 800,000 Hawaiians lived in a society with highly complex political and social systems. Separate high chiefs governed the major islands, with subordinate chiefs managing self-sustaining land units encompassing broad plains near the sea, running up valley ridges to the mountains. Within these units, the people used the resources necessary to sustain life—access to offshore fishing and shoreline gathering; plots of land and sufficient water for growing taro, banana, breadfruit, or sweet potatoes; the right of way to the uplands for timber and fuel; and the right to hunt and gather wild plants and herbs.

This would all change as a result of contact with Europeans. On his third voyage into the Pacific, the explorer Captain James Cook, British commander of HMS Resolution and HMS Discovery, on January 18, 1778 found Oahu and Kauai. He was thought of by the Hawaiians as the reincarnation of Lono, one of their principal gods. Cook named Hawaii the Sandwich Islands in honor of the Earl of Sandwich. He returned to Hawaii a year later and was slain there on February 13, 1779. As a result of the European contact, contagious diseases including cholera, measles and gonorrhea, decimated the Hawaiian population. The population was estimated at between 250,000 to 1 million when Captain Cook sailed into Kealakakua in 1779. By 1848 Hawaiians numbered 88,000. By the year 2044, demographers predict that there will be no more Native Hawaiians left.

Figure 23.3. Captain James Cook.

23.3   BACKGROUND

Many Native Hawaiians share the view that tourism, as a foreigner dominated enterprise, is the plague which an already oppressed people must endure with very few other economic options or alternatives in life. Still, many end up choosing the lesser options even if it means unemployment or criminal activity. It is no accident that Hawaiians are the poorest of all people in Hawaii, with the highest percentage of unemployment, welfare recipients and prison populations. Notwithstanding the tourist industry, Native Hawaiians continue to be the poorest, sickest and least educated of all people in Hawaii.

Tourism is concerned with self-preservation as an industry and not with the well-being of the community. In March of 1991, during the dramatic decline of visitors to Hawaii due to the Gulf War, the Hawaii State Legislature readily allocated as an emergency measure $6 million to be used by the Hawaii Visitors Bureau for television commercials on the mainland USA. During the same period, hundred of hotel employees were laid off in one of the largest layoffs in recent years. No emergency measures to assist the unemployed were introduced or even considered.

Foreign investment related to tourism went from 70.8 million dollars in 1981 to over a billion and a half in 1986. Japanese investment is over 3 billion dollars for hotels alone from 1989-2007. The Australians remain in second place with 117 million invested. Today, almost every major hotel is owned by foreign investors and nearly every hotel planned is being funded by foreign investment.

Business interests caused the illegal overthrow of the Hawaii Nation in 1893. A hundred years later, the same business driven interests continue and Native Hawaiians remain victims of an exploitation whose guise is an industry called tourism. A basic human right is the ability of a people to be self-governing, self-determining and self-sufficient. This right was taken away from Hawaiians when the nation was overthrown. Tourism in many respects perpetuates the oppression.

In its current form, tourism has evolved to a point where it is of minimal economic consequence to Native Hawaiians. Tourism, which stifles economic diversification and weakens existing agricultural and technological development, does not provide a viable economic alternative to Native Hawaiians.

It is interesting to consider the increase in the number of hotel rooms since 1985 (Fig. 23.4). Along with this data, consider the increases observed in actual and projected visitor numbers (Fig. 23.5).

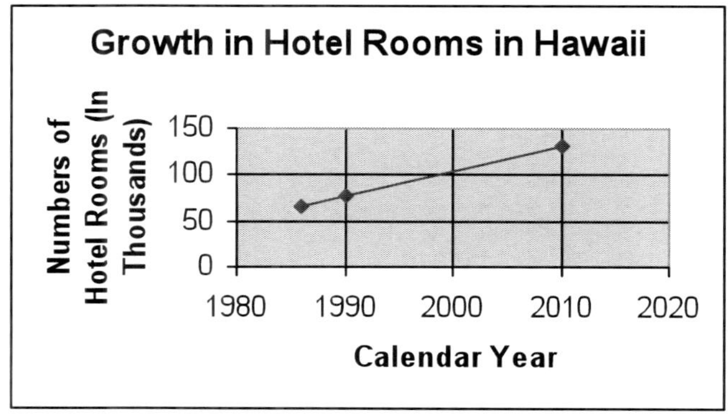

Figure 23.4. Growth in Numbers of Hotel Rooms in Hawaii since 1986.

Hawaii, it seems, is headed toward a non-diversified economic future that will be totally dependent upon the tourist industry. Tourism therefore, will continue to be a major obstacle in the movement of Hawaiians towards self-governance and self-determination. More poverty and the continuation of the negative impact upon Native Hawaiians will be the inevitable outcome of the future of tourism in Hawaii.

When the primary means of promotion is dependent upon a culture and people, and the perception that "all is well in paradise" is put forward while in fact "all is not well," then the issue becomes one of cultural prostitution. It becomes the selling of an artificial cultural image that has complete disregard for the truth, at the expense and pain of Native Hawaiians who are struggling

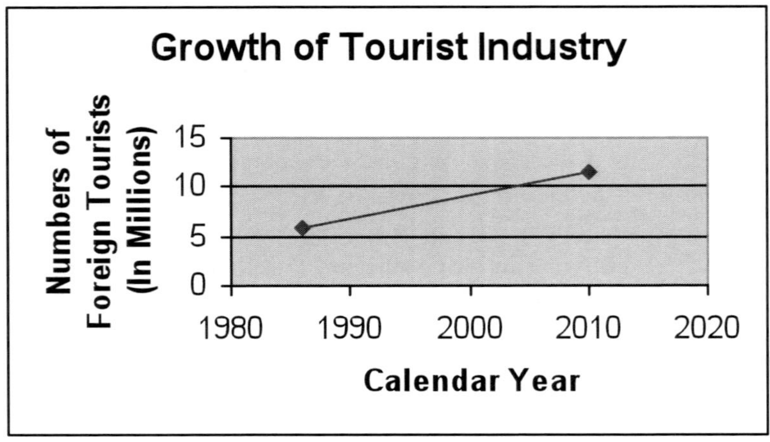

Figure 23.5. Increase of Foreign Visitors to Hawaii since 1986.

to survive. From printed brochure to life in the fast lane, tourism promotes the development and practice of an entertainment and visitor oriented culture. The follow-through with marketing and promotion is part and parcel of the "plastic tikis, Kodak hula, and concrete waterfalls."

Tourism development in Hawaii most often takes place at the expense of a people's cultural and historical symbols and land based resources. Tourism development has played a major role in the destruction of ancient Hawaiian burial grounds, significant archaeological historic sites and sacred places. Almost every major resort development has been built on some culturally significant site. Community opposition is usually based upon these cultural issues. The usually insensitive approach and manner of development leaves the local community to conclude that there is no respect or concern for the culture and identity of Hawaiian people.

The third major impact of tourism on Native Hawaiians must be understood in the context of environmental exploitation. The character of indigenous Pacific cultures in relationship to the land is one based on a high level of environmental awareness and ecological conservatism. The relationship of people to land, and people to sea, is spiritual and religious. Land is the base around which a culture evolves. When tourism takes away the land, takes away access to the fishing grounds or the right to gather food or medicine, the Hawaiian loses a primary means of livelihood, and more importantly, meaning in life. Crowded beaches and commercial tour boating do much to threaten shoreline or coastal fishing through noise or chemical pollution. The state has begun to identify beach parks and near-shore areas that are exceeding capacity use because of significant resident and visitor numbers. User conflicts between residents and visitors are becoming a problem and are expected to escalate as tourism and ocean recreation industries continue to grow.

There is a growing frustration among Native Hawaiians. This expresses itself in the very visible opposition to resort of related development and the increasing amount of land being openly occupied by indignant Hawaiians. Almost every large resort development in the last ten years has been opposed by Hawaiian groups or organizations.

The emerging awareness of the impact of tourism on Native Hawaiians captured local and international attention when, in August 1989, an international conference on tourism sponsored by local, national and international church groups and organizations took place in Hawaii. That conference looked at the negative impact of tourism on Native Hawaiians. The results were published in what came to be called the Hawaiian Declaration of 1989. The story of the Native Hawaiian people, a people who love their land, is a complicated and difficult one. But when told in broad strokes, it is a familiar one: a story of an indigenous people and of greed, racism, and imperialism.

23.4   CASE STUDY

Scenario: Market analysis has shown that there will be a huge increase in tourism and foreign visitors to Hawaii. This increase is in part due to the threat of terrorism in Europe and the Middle East. The airport at Honolulu is in need of significant improvements in its runways, which will require the acquisition of large tracts of land adjacent to the existing airport and several large parcels at other locations in the city. You are a design engineer employed by the multi-national engineering firm, which won the contract for the renovations and expansions. Your firm is world renowned for its strict adherence to principles of green design.

Approach: Using the various decision making models described in Book 2 and in the last case study, let us explore the kinds of questions that such models challenge us to include in our deliberations and ultimately our design.

- Engineering and Freedom

Consider first an engineering decision model based upon freedom. We considered three kinds of freedom: existential freedom, substantial freedom and the freedom of choices.

So what kinds of questions should we consider in our airport renovation and expansion design deliberations? Should we consider the impacts of increased tourism? Do we consider the effects on the environment? The native populations? Their history? Their culture?

- Engineering and Chaos

Engineering decisions using a chaos perspective caution us to focus on the health of the various ecosystems and the natural world in general. What are the impacts of the increase in tourism and foreign visitors on the local ecosystems? The health of the planet? What are the longer term impacts? What effects will the increase in foreign visitors have on the various cultures?

- Engineering and a Morally Deep World

The fundamental difference between an ethical code based on a morally deep world versus our present sense of responsibility is the replacement of the "public" by the "identified integral community." Who then is included in the integral community?

- o Multinational corporations? Their shareholders? Their employees?

- o Foreign visitors?

- o Native Hawaiians?

- o Local ecosystems? Endangered plants and animals?

- o Anyone else?

- Engineering and Globalism

As we discussed earlier, globalisation, in its literal sense, is the process of making, transformation of some things or phenomena into global ones. The question that we are asked to consider is the following: Is the increase in tourism and foreign visitors consistent or in opposition to the *Global Ethic*? By *Global Ethic* is meant the necessary minimum of common values, standards and basic attitudes or alternatively a minimal basic consensus relating to binding values, irrevocable standards and moral attitudes.

- Engineering and Love

In order to understand the nature of an engineering based on love, we sought wisdom from Thomas Berry. According to Berry, our new community is a very special one; that is, it is one in which the various elements are bound together as subjects having interests rather than one in which some have interests while others are simply resources to be utilized. What are the implications then for our airport renovation and expansion project? Have we considered all those who will be impacted as objects or as subjects? Are they simply cogs in our wheels of career advancement? Economic development? And what of the various ecosystems? How will they be impacted?

Bibliography

Abdala, A., pp. 285–310, Reunión Anual, 32, Bahia Blanca, 19-21 Noviembre 1997; Bahia Blanca, AAEP, 1997.

AIChE Code of Ethics, 2008.
http://www.stuorg.iastate.edu/aiche/ethics.html

Aiello, R. and Grajales, F., Social aspects of Solid Waste management: The experience of Argentina, World bank Urban Development website, World Bank Group 2001.

Alcazar, Lorena; Abdala, Manuel A.; Shirley, Mary M., The Buenos Aires Water Concession, Volume 1 World Bank Report number WPS2311, 2000.

Allenby, B., *Micro and Macro Ethics for an Anthropogenic Earth*, Professional Ethics Report, Vol. 18, No. 2, Spring 2005.

Amartya Sen, *Development as Freedom*, Anchor Press, 2005.

Annas, J., *Virtue Ethics*,
http://www.u.arizona.edu/~jannas/forth/coppvirtue.htm

Asbestos Hazards Handbook - Chapter 3, London Hazards Centre, 1995.

ASCE Code of Ethics, 2008.
http://www.asce.org/inside/codeofethics.cfm

Asch, S., *Social psychology.* Englewood Cliffs, NJ: Prentice-Hall, 1952.

ASME Code of Ethics, 2008.
http://courses.cs.vt.edu/~cs3604/lib/WorldCodes/ASME.html

Asprey, R., *The Rise of Napoleon Buonoparte*, Basic Books, 2002.

Aurelius, Marcus, Hicks, D., and Hicks, C. Scot, *The Emperor's Handbook: A New Translation of the Meditations*, Scribner, 2002.

Austin, J., *The Province of Jurisprudence Determined*, Cambridge University Press, 1995.

Baillie, C., 'The Travelling Case: How to foster Creative Thinking in Higher Education,' Published by the Higher Education Funding Council, England, 2003.

Baillie, C., Engineers within a local and global society: Synthesis lectures on engineers, technology and society, series editor, Morgan and Claypool, US 2006.

Baillie, C., Learning to learn, Imperial College, 1998.

Bak, Per, How Nature Works: The Science of Self-Organized Criticality, Copernicus Books, 1996.

Barker, A. and Peters, G., Eds. The politics of expert advice : creating, using and manipulating scientific knowledge for public policy Pittsburgh, Pa., University of Pittsburgh Press, 1993.

Barker, Jonathan, Street-level democracy: political settings at the margins of global power, Between the Lines, Canada, 1999.

Barlow, M., Bluegold: The fight to stop the Corporate theft of the Worlds water, Newpress, 2003.

Barlow, M., 2007.

Barlow, M., 2008.

Bernal, J., Towards a theory of Innovation in Services, Research Policy 15, no 4, p161–73, 1939.

Berry, T., *The Dream of the Earth,* Sierra Club, October 2006.

Bhattachary, D., Sheppard, C., and Boak, D., Royal society (UK), Science in Society, Report 2004.

Biggart, John, Glovelli,Georgii, and Yassour, Avraham, Alexander Bogdanov and the Origins of Systems Thinking in Russia. Avebury, 1998.

Bijker, W., Hughes, T. P., Pinch, T., Eds., The Social Construction of Technological Systems: New Directions in the Sociology and History of Technology, MIT press, 1998.

Bijker, W., Of Bicycles, Brakelites and Bulbs: Towards a theory of sociotechnical Change, Cambridge, MA, MIT Press, 1995.

Bogdanov, A., The Struggle for Viability, Xlibris Corporation, 2002.

Botkin, Daniel, Discordant Harmonies, Oxford University Press: New York, 1993.

Brainy Quotes, 2008.
http://www.brainyquote.com/quotes/quotes/e/eowilson164833.html

Brawley, M., The Politics of Globalisation, Broadview Press, 2003.

British Asbestos Newsletter Issue 26, Winter 1996/7.

British Asbestos Newsletter Issue 37, Winter 1999/2000.

Bruchhaus, E., Improved cooking stoves: miracle weapon in the fight against the desert? Development and cooperation 1/85: 24-26, 1985.

Brunner, R., Steelman, T., Coe-juell, L., Cromley, C., Edwards, C., and Tucker, D., Adaptive governance : integrating science, policy, and decision making New York : Columbia University Press, 2005.

Bunge, Mario, Causality: The place of the causal principle in modern science, World Publishing Co., 1963.

Capra, F., The Hidden Connections: a science for sustainable living, Anchor Books, 2002.

Carrada, G., Communicating science: a scientists survival kit, Italian conference of Deans, report 2005.

CEAMSE, Estudio de calidad de los residuos solidos urbanos, CEAMSE, Enero 2007.

Chomsky, N., "Education is Ignorance," excerpted from *Class Warfare*, 1995, pp. 19–23, 27–31.

Coburn, J., Street science : community knowledge and environmental health justice, Cambridge, MA : MIT Press, 2005.

COCD, Creaddenda – Deskundigheids opleiding, COCD, Antwerp, 1994.

Collins, H. and Evans, R., The third wave of science studies: studies of expertise and experience' Social studies of science 32(2) 235-296 2002.

Common Dream.org, 0000.
http://www.commondreams.org/views05/0907-26.htm

Cook, M., *The Moral Warrior: Ethics and Service in the U.S. Military*, Albany, NY, State University of New York Press, 2004.

Cook, Steve, "More Americans Breathing Dirty Air, Lung Association Says in 2001 Report," Daily Report for Executives, Bureau of National Affairs. May 2, 2001. Page A-11.

Cowen, M. P. and Shenton, R. W., Doctrines of Development, Routledge, 1996.

Creighton, J., The public participation handbook : making better decisions through citizen involvement, 1st ed., San Francisco : Jossey-Bass, 2005.

Dalai Lama and Cutler, H., *The Art of Happiness: A Handbook for Living*, Riverhead Press, 1998.

Daly, H., Population, Migration, and Globalization, Worldwatch Magazine,
http://www.publicpolicy.umd.edu/faculty/daly/WW%20rev%20pop,migr, glob%20copy%201.pdf

Davis, M., *Thinking Like an Engineer: Studies in the Ethics of a Profession,* Oxford University Press, 1998.

Davis, M., *Ethics and the University*, Routledge, 1999.

Davis, M., Late Victorian Holocausts: El Nino Famines and the making of the Third World, Verso Press, 2002.

Decker, 1995.

http://www.deephawaii.com/hawaiianhistory.htm

Dewulf, S. and Baillie, C., CASE: Creativity in Art, Science and Engineering, DFEE, UK, 1999.

Drori, G., A critical appraisal of science education for economic development, Chapter 3 p 49-74 in William Cobern, Socio cultural perspectives on science education: an international dialogue, Science and Education Library, Kluwer, 1998.

Effectiveness and Impact of Corporate Average Fuel Economy Standards. Division on Engineering and Physical Sciences, Board on Energy and Environmental Systems, Transportation Resource Board. National Research Council, July 2001. Page ES-4.

Ellis, J., *Colony Collapse Disorder (CCD) in Honey Bees,* http://pestalert.ifas.ufl.edu/Colony_Collapse_Disorder.htm

Ellul, J., The Technological Society, Vintage Press, 1967.

Engels, F., Conditions of the working class in England, Harmondsworth: Penguin, 2004.

Environmental Justice in Hawaii: *A Hawaiian Issue,* Rev. Kaleo Patterson, http://members.tripod.com/~MPHAWAII/Tourism/ EnvironmentalJusticeInHawaii.htm

EPA Assessment Factors, Science Policy Council, United States Environmental Protection Agency EPA 100/B-03/001 June 2003.

Francis, R., Water justice in South Africa: Natural Resources Policy at the intersection of Human Rights, Economics and Political Power, Berkely Electronic Press (law.bepress/.com/expresson/eps/518) 2005.

Franklin, U., The Real World of technology, House of Anansi Press, 1990.

Frayn, M., Copenhagen, Anchor Books 1998.

Friedman, T., *The World is Flat 3.0*, Picador Press, 2007.

Fuel Economy website. Department of Energy and Environmental Protection Agency (`http://www.fueleconomy.gov/feg/climate.shtml`)

Furslang, L., Three perspectives in STS in the Policy Context, in 'Visions of STS,' counterpoints in science, technology and society studies, Cutcliffe, H. and Mitcham, C., Eds., p35–47, 2001.

Galiani, S., Gertler, P., Schargrodsky, E., Water for life: The impact of the privatization of water services on child mortality, Journal of Political Economy, Vol. 113, pp. 83–120, February 2005.

Garrett, J., *Amartya Sen's Ethics of Substantial Freedom*, `http://www.wku.edu/~jan.garrett/ethics/senethic.htm`

Gelb, M.J., Putting your Creative Genius to Work: How to Sharpen and Intensify your Mind Power, Nightingale Conant, Illinois, 1996.

Gelinas, N., *Katrina's Real Lesson*, City Journal, Vol. 17., No. 1, Winter 2007.

Gobierno de la Ciudad de Buenos Aires Equipo Tecnico del Plan Estrategico, Indicatores, Part 2, Version Preliminar, marzo, p3. 1998.

Graham, L., What have we learned about science and technology from the Russian experience? Stanford University Press 1998.

Graves, K., Golden Rule by Sextus, *The World's Sixteen Crucified Saviors*, Colby and Rich, 1876.

Greene, Brian, The Elegant Universe: Superstrings, Hidden Dimensions, and the Quest for the Ultimate Theory, Norton, 1999.

Gupta, A., Science, sustainability and social purpose: barriers to effective articulation, dialogue and utilization of formal and informal science in public policy International Journal of Sustainable Development, Vol. 2, No 3, 1999.

H M Treasury Pre Budget report: Investing in Britain's potential Building our long term future, Cm6984 Dec 2006.

Hampson, S., Trying to know the unknown, The Globe and Mail. Jan 17th, R3, 2004.

Hansen, J., D. Johnson, A. Lacis, S. Lebedeff, P. Lee, D. Rind, and G. Russell, 1981: Climate impact of increasing atmospheric carbon dioxide. Science, 213, 957-966, doi:10.1126/science.213.4511.957.

Hargreaves, I., Lewis, J., and Speers, T., Towards a better map, Science the public and the media, ESRC report, 2003.

Harris, C., Pritchard, M., and Rabins, M., *Engineering Ethics: Concepts and Cases*, Wadsworth Pub., July 2008.

Hawking, Stephen, A Brief History of Time, Bantam Press: New York, 1998.

Herkert, J., *Engineering Ethics and Climate Change*, ASEE 2008, Pittsburgh, Pa., June 2008.

Hevesi, D., Hasidic and Hispanic residents in Williamsburg try to forge a new unity, New York Times sept. 18 B47, 1994.

Hinman, L.,
`http://ethics.sandiego.edu/presentations/Theory/Utilitarianism///index_files/v3_document.html`

Holmes, 2002.
`http://users.ece.utexas.edu/~holmes/Teaching/EE302/Slides/UnitOne/sld002.htm`

Hounshell, David, Assembly Line, Johns Hopkins Univeersity Press: Baltimore, MA 1984.

"Hurricane Katrina,"
`http://en.wikipedia.org/wiki/Hurricane_Katrina`

Hurricane Katrina Damage- Louisiana, Adjutant General's Department State of Kansas,
`http://www.accesskansas.org/ksadjutantgeneral/Disaster-Emergency/2005/Hurricane%20Katrina/Damage/Damage%20-%20Louisiana.htm`

IEE Code of Ethics, 2008.
`http://www.iienet2.org/Landing.aspx?ID=299`

IEEE Code of Ethics, 2008.
`http://www.ieee.org/portal/pages/iportals/aboutus/ethics/code.html`

Imber, J., Ed., Searching for science policy, New Brunswick, NJ: Transaction Publishers, 2002.

Intergovernamental Panel on Climate Change (IPCC), 2001.

Irwin, A., Citizen Science: A study of people, expertise and sustainable development, Routledge 1995.

Isaksen, S.G., Murdock, M.C., Firestien, R.L., Eds., Nurturing and Developing Creativity: The Emerge of a Discipline, Alex Publishing Corporation, New Jersey, 1993.

Jagger, B., Cluster Bombs and Teddy Bears, Telegraph.co.uk., May 11, 2007,
`http://www.telegraph.co.uk/opinion/main.jhtml?xml=/opinion/2007/11/05/do0502.xml`

James, Jamie, The Music of the Spheres: Music, Science, and the Natural Order of the Universe, Copernicus Books, 1993.

Jefferson, T., to Maria Cosway, 1786. ME 5:443.

Johnson, L., *A Morally Deep World*, Cambridge University Press, 1993.

Johnson, L., 1991.

Kanbur, R., Development disagreements and water privatization: bridging the divide, `www.people/cornell/pages/sk145/papers.htm`) 2007.

Kant, I., 'Preface.' In *The Metaphysical Elements of Ethics*. Translated by Thomas Kingsmill Abbott, 1780.

Kashner, A., Professional Ethics and Collective Personal Autonomy, *Ethical Perspectives*, 12/1, March 2005, pp. 67–97.

Kass, G., Office of science and innovation, UK, Guiding principles for public dialogue, Sept 2006.

King, M.L., and Washington, J.M., *A Testament of Hope: The Essential Writings and Speeches of Martin Luther King*, Harper One, 1990.

Kipphardt, H., In the matter of J. Robert Oppenheimer. Trans Ruth Speirs, New York: Hill and Wang, 1968.

Kirkpatrick, C., Parker, D., and Zhang, Y.-F., State versus private sector provision of water services in Africa: an empirical analysis, Presentation Centre on Regulation and Competition, 3rd International conference, Pro-poor regulation and competition: Issues, Policies and Practices, Cape Town, South Africa, Sept 2004.

Kuhn, T. S., The structure of scientific revolutions. Chicago: University of Chicago Press, 1962.

Landmines: Africa's Stake, Global Initiatives, `http://www.africaaction.org/bp/lmineall.htm`

Leach, M., Scoones I., and Wynne, B., Eds., Science and citizens : globalization and the challenge of engagement. London ; New York : Zed Books ; New York : Distributed in the USA exclusively by Palgrave Macmillan, 2005.

Leftwich, A., States of Development, Polity Press, 2000.

Leopold, A., *Sand County Almanac*, Oxford University Press, 1968.

Liaschenko, J., Oguz, N. and Brunnquell, D., *Critique of the Tragic Case Method in Ethics Education, Journal of Medical Ethics*, 32(11):672.

Light-Duty Automotive Technology and Fuel Economy Trends 1975 Through 2000. Executive Summary. Robert M. Heavenrich and Karl H. Hellman. Advanced Technology Division, Office of Transportation and Air Quality. U.S. Environmental Protection Agency. Air and Radiation EPA420-S-00-003 (`http://www.epa.gov/otaq/fetrends.htm` 06 Jan 2001).

Loftus, A. and McDonald, D., Lessons from Argentina: The Buenos Aires Water Concession, Municipal Services project Occasional Papers Series Number 2, 2001.

Luke 6, 27–32, English Standard Version Bible.

MacColl, E., Journeyman: An autobiography by Ewan MacColl. London: Sidgwick and Jackson, 1990.

MacFarquhar, L., A dry soul is best: Friendship, espionage and the plays of Michael Frayn, New Yorker, Oct 25, p64-73, 2004.

MacKenzie D. and J. Wajcman, Eds. The Social Shaping of Technology. Philadelphia: Open University Press, 1985.

Marsh, G.P., *Man and Nature*, Kessenger Publishing, 2008.

Marshall, Alfred, The Principles of Economics, Prometheus Books, 1997.

Marx, K., Fernbach, D. and Mandel, E., Capital: a critique of political economy, Harmondsworth: Penguin, 2004.

Mastny, Lisa, Ed., Vital Signs 2006, W.W. Norton & Co., 2005.

Mayer-Kress, G., *What is Chaos?*, Sat Apr 22, 21:04:59 MDT, 1995,
http://www.santafe.edu/~gmk/MFGB/node2.html

McDonald, D. and Rulters, G., Rethinking privatization: Towards a critical theoretical perspective, Public Services yearbook, Part 1, Chapter 1, 2005/6.

Medvitz, A.G., Problems in the application of science education to national development Nairobi, Kenya, Institute of development Studies, University of Nairobi, 1985.

Menzies, H., Whose Brave New World: The Information Highway and the New Economy. Toronto: Between the Lines, 1996.

Metairie, Louisiana, Wikipedia,
http://en.wikipedia.org/wiki/Metairie%2C_Louisiana

MFGB, 0000.
http://www.santafe.edu/~gmk/MFGB/node2.html

Mighton, J., Scientific Americans, Playwrights Canada Press, Toronto, 1987.

Miller, J. and Valdez, W., Science policy leaders in the United States Presented at the annual meeting of the American Association for the Advancement of Science, Denver Colorado, 2003.

Mills, J.S., *On Liberty and Other Essays*, Oxford University Press, 2008.

Morris, M. and Muzychka, M., Participatory research and action research: A guide to becoming a researcher for social change: CRISW-ICREF (Canadian Research Institute for the Advancement of Women) 2002.

Motor Vehicle Facts and Figures, 1997. American Automobile Manufacturers Association. p. 84.

Muller, H., The Children of Frankenstein: A Primer on Modern Technology and Human Values. Bloomington & London: Indiana University Press, 1971.

Mumford, L., Technics and civilization. New York,: Harcourt Brace & World, 1963.

Mumford, L., The myth of the machine: technics and human development, 1st ed. New York: Harcourt Brace & World, 1967.

Mumford, Lewis, Technics and Civilization, Harvest/HBJ Books: New York, 1963.

Murphy, G., *Human Potentialities*. New York: Basic Books, 1958.

Mutnick, D., Critical interventions: the meaning of praxis, in A Boal companion: Dialogues on theatre and cultural politics, Cohen-Cruz, J. and Schtzman, M., Eds., Routledge, p33-46, 2006.

Nature editorial, 2003, "Welcome to the Anthropocene," *Nature* 424:709.

Naughton, Keith, "The Unstoppable SUV," Newsweek, July 2, 2001.

News Hounds,
http://www.newshounds.us/2005/08/31/when_the_levee_breaks.php

NOAA, 2005.
http://www.katrina.noaa.gov/satellite/images/katrina-08-29-2005-1415z.jpg

Noble, D., America By Design: Science, Technology and the Rise of Corporate Capitalism. Oxford: Oxford University Press, 1977.

Noble, D., Forces of Production: A Social History of Industrial Automation. Oxford: Oxford University Press, 1984.

NSPE Code of Ethics, 2008.
http://www.nspe.org/Ethics/CodeofEthics/index.html

Nussbaum, M.C., *Sex and Social Justice*, Oxford University Press, 1999.

Oates, S., Owen, D., and Gibson, R., Eds., The Internet and Politics: Citizens, voters and activists. Routledge, London and NY, 2006.

Osborne and Van Loon, 1998.

Osborne, R., and Van Loon, B., Introducing Sociology, Icon Books, 1999.

Pacific Disaster management Information, Network 2004.

Panch, T., *Enduring Effects of War: Health in Iraq 2004*, Medact, London, UK, 2004.

Parliamentary Office of Science and Technology, UK, Open Channels: Public dialogue in science and technology, Report no 153, 2001.

Parry News Issue no 47, May 2007.

Peppard, M., Friedrich Durrenmatt, New York, Twayne, 1969.

Petroski, H., *To Engineer is Human: the Role of Failure in Successful Design*, St Martins Press, 1985.

Pirsig, R., Zen and the Art of Motorcycle Maintenance Vintage, p196, 1989.

Pojman, L.P., and Fieser, J., *Ethics: Discovering Right and Wrong*, Wadsworth Cengage Learning, Wadsworth: Belmont, Ca., 2009.

Polanyi, K., The Great Transformation: The political and economic origins of our time, Beacon Press, 2001.

POSTnote Parliamentary office of Science and Technology, UK, Nov Number 189, 2002.

Powers, T., The unanswered questions, New York Review of Books, 47, 9, May 25th 2000.

Quoteworld, 2008.
http://www.quoteworld.org/category/nature/author/ralph-waldo-emerson

Reader, J., Globalisation, Engineering and creativity: Synthesis lectures on engineers, technology and society, series editor, Morgan and Claypool, US 2006.

Roberts, Paul, "Bad Sports," Harper's Magazine. April 2001.

Salverson, J., Witnessing subjects: a fool's help, in A Boal companion: Dialogues on theatre and cultural politics, Cohen-Cruz, J. and Schtzman, M., Eds., Routledge, p146-159, 2006.

Samson, M., Dumping on women: Gender and privatization of waste management, SAMWU (South African Municipal Workers Union and Municipal Services project), 2003.

ScienceDaily, Apr. 23, 2007.

Sclove, R., Democracy and Technology, Guilford, NY, 1995.

Sen, A., Development as Freedom, Anchor Books, 1999.

Shapin, S., 'Here and everywhere: sociology of scientific knowledge.' Annual Review of Sociology, 21, 289-333, 1984.

Shepherd-Barr, K., Science on stage: From Dr Faustus to Copenhagen, Princeton University Press, 2006.

Sherman, W.T., *Memoirs*, Penguin Classics, 2000.

Sim, S. and Van Loon, B., *Introducing Critical Theory,* Icon Books, Duxford, UK, p. 165, 2000.

Sim., S. and Van Loon, B., Introducing Critical Theory, Icon Books, 2001.

Simon, S.B., Howe, L.W., and Kirschenbaum, H., Values Clarification, Grand Central Publishing; Revised edition (September 1, 1995).

Singer, P., *In Defense of Animals: The Second Wave*, Wiley-Blackwell, 2005.

Singh, G.B., *Gandhi: Behind the Mask of Divinity*, Prometheus Books, 2004.

Sismundo, S., An introduction to Science and Technology Studies, Blackwell, 2004.

Smillie, I., Mastering the machine: Poverty, Aid and Technology, Westview Press, Colorado, 1991.

Smith, Adam, The Wealth of Nations, Bantam Classics: New York, 2003.

Smith, H.W., *Aeschylus*, Harvard University Press, 1970.

Stanford Encyclopedia of Philosophy,
  http://plato.stanford.edu/entries/rights/

Stephenson, S., An experiment with an airpump, London, Methuen, 1998.

Stewart, I., *Does God Play Dice*, Wiley Blackwell, 2002.

Stewart, I., 1998.

Stiglitz, J., Globalisation and its discontents, Norton and company, 2003.

Stilgoe, J. and Warburton, D., Science horizons shaping our future: Broadening our horizons – public engagement with the future of science 2006.

Stockdale, J., *A Vietnam Experience: 10 Years of Reflection*, Hoover Institution Press, 2004.

"Summary for Policymakers: A Report of Working Group I of the IPCC, 2001."
  (http://www.ipcc.ch/)

Swift, J. and Stewart, K., Hydro: the decline and fall of Ontario's electric empire, Between the Lines, Canada, 2004.

The Universality of the Golden Rule in the World Religions,
   www.teachingvalues.com/goldenrule.html

Tennyson, The Works of Tennyson, Bibliolife, 2009.

Tesconi, T., "California Grape Harvest 2000," *The Press Democrat*, October 26, 2000.

The 1989 Hawai's Declaration of the Hawai's Ecumenical Coalition on Tourism Conference.

The Stanford Dictionary of Philosophy, 2009.

The Ten Commandments, Catholic Encyclopedia,
   http://www.newadvent.org/cathen/04153a.htm

The Worldwatch Institute, State of the World 2006, Worldwatch Institute, 2006.

Tolle, Eckhart, The Power of Now, New World Library: New York, 2004.

Tomic, W. and Brouwers, W., Idea Generating Among Secondary School Teachers. The Open
   University (unpublished) Heerlen, The Netherlands, 1998.
   TOURISM IN HAWAI'I: Its Impact on Native Hawaiians And Its Challenge to the Churches,
   http://members.tripod.com/~MPHAWAII/Tourism/The1989.htm

Tourisms Negative Impact on Native Hawaiians, Rev. Kaleo Patterson,
   http://members.tripod.com/~MPHAWAII/Tourism/TourismsNegativeImpact.htm

Turkle, S., Identity in the age of the internet, Life on the screen, Simon and Schuster, NY, 1995.

Twain, M., "To The Person Sitting in Darkness," Mark Twain's Weapons of Satire: Anti-Imperialist
   Writings on the Philippine-American War, Syracuse University Press, 1992.

Ulam, S., on Chaos, 0000.
   http://www.santafe.edu/~gmk/MFGB/node2.html

UNESCO, International Tsunami Information Center, 2005,
   http://ioc3.unesco.org/itic/

Vanderburg, W., The Labyrinth of Technology, University of Toronto press, 2000.

Vanderburg, W., The Development of Minds and Cultures, University of Toronto Press, 1985.

von Clausewitz, Carl, *On War*, Eco Library, 2007.

Warner, C., "All Mankind Is One:" The Libertarian Tradition In Sixteenth Century Spain. *The
   Journal of Libertarian Studies*, 8, 2:293–309, 1987.

Watts, Stephen, The Peoples Tycoon: Henry Ford and the American Century, Knopf Press, 2005.

Random House Webster's Unabridged Dictionary, Random House Reference, 2005.

White House: U.S. will ban some land mines, CNN.com, February 27, 2004,
http://www.cnn.com/2004/US/02/27/bush.landmines.ap/

Whitmarsh, L., Kean, S., Russell, C., and Peacock, M., Connecting Science: what we know and what we don't know about science in society, British Association for the Advancement of Science, 2005.

Wickham, G., Drama in a world of science, in Drama in a world of science and three other lectures, London:Routledge and Kegan Paul, 1962.

Wikipedia, 2008.
http://en.wikipedia.org/wiki/Brian_Swimme

Wolfe, Maynard, Rube Goldberg Inventions, Simon and Schuster, 2000.

World Bank 1999: World Development report 1998/1999: Knowledge for development.
http://econ.worldbank.org/wdr/. 1999.

World Bank Development indicators, World Bank webpage, table 3.10 Urbanisation, 1999.

Worldwatch Institute, *State of the World 2007*, New York: W.W. Norton, 2007.

Yeger, S., The Sound of one hand clapping: A guide to writing for the theatre, Amber lane press, Oxford, 1990.

Zuckerman, H., 'The sociology of science.' In N. J. Smelser, Ed., Handbook, 1988.